MW00723683

reach for the
STARS

reach for the STARS

The Evolution of India's Rocket Programme

GOPAL RAJ

VIKING

VIKING
Penguin Books India (P) Ltd., 11 Community Centre, Panchsheel Park, New Delhi 110017, India
Penguin Books Ltd., 27 Wrights Lane, London W8 5TZ, UK
Penguin Putnam Inc., 375 Hudson Street, New York; NY 10014, USA
Penguin Books Australia Ltd., Ringwood, Victoria, Australia
Penguin Books Canada Ltd., 10 Alcorn Avenue, Suite 300, Toronto, Ontario M4V 3B2, Canada
Penguin Books (NZ) Ltd., Cnr Rosedale & Airborne Roads, Albany, Auckland, New Zealand

First published in Viking by Penguin Books India 2000

10 9 8 7 6 5 4 3 2 1

Printed at Rekha Printers Pvt. Ltd., New Delhi

Contents

Preface

THIS BOOK IS not intended as a history of the Indian launch vehicle programme in the sense that I have made no attempt to comprehensively cover all the developments carried out by the Indian Space Research Organization (ISRO) in this field. Rather, my aim has been to look at how the Indian launch vehicle programme evolved, why it took the path that it did and the strengths and weaknesses which this trajectory has created.

The Indian launch vehicle programme, and indeed its space programme, has been unique in many ways. Several developing countries have at various times entertained ambitions of having a space programme and developing an independent launch capability. But China and India are the only developing countries to have successfully done so. In the case of the Chinese, however, security threats from the then Soviet Union as well as the United States drove them to establish an atomic bomb capability and to develop long-range ballistic missiles to deliver those nuclear warheads. These ballistic missiles and the experience gained in developing them became the basis of the Chinese Long March launcher programme.

The Indian launch vehicle programme, on the other hand, had to be built from scratch. This involved ensuring sustained funding from the government, developing the technological capability to build launch vehicles and coping with the limited industrial capability available in a developing country. The greatest good fortune for the Indian space programme, as I will argue in this book, was the vision of its founder, Vikram Sarabhai. He foresaw the applications which would most benefit India and decided that the country ought to have the capability to build satellites to meet those applications as well as the launchers to put the satellites into orbit. It was Sarabhai's emphasis on applications, an emphasis continued by his successors, which has ensured continued support from the government.

There is nothing inevitable about the path which ISRO has followed in the development of its launch vehicles. At every step along the way, there were alternative technology choices possible. A number of factors then influenced the course of events. Technical capability which has already been developed and which can therefore be readily built upon represents an important consideration. On the other hand, the development of completely new technologies may be advantageous. At the management level, technology choice has time and cost implications which have to be factored in. There can be issues of whether to develop the technology indigenously or import it. Ultimately, technology choices are not matters decided with clinical detachment. These are issues over which competing groups and individuals fight grim battles, looking for the funding and opportunities for growth and advancement.

In present-day launch vehicles, the propulsion systems employed have an overwhelming influence on their configuration. The reason is not hard to see. Propellants alone account for 75 to 80 per cent of a launcher's lift-off weight while the payload they can carry to orbits close to earth will only be two to three per cent of their weight. Contrast this with a modern Boeing 747 cargo aircraft. Fuel takes up only about 40 per cent of its maximum take-off weight while 30 per cent of the weight is cargo. Consequently, in this book too I have concentrated heavily on technology options, choices and development in propulsion systems for ISRO's launch vehicles.

There is no question that launch vehicle development is much more than just propulsion systems governed by a guidance package. It is a truly multi-disciplinary team effort, requiring expertise in such diverse areas as aerodynamics, structures, materials, electronics, chemicals, mechanical engineering, composites, computers and software development, pyrotechnic devices and so on. But I felt that to examine technology development in each of these areas would take the readers into dense thickets of technical detail. Such an approach would not serve my purpose of providing a coherent overview of the Indian launch vehicle programme.

Over the years, as I have studied the space programme, I have got to know not just the technology, but also the people and the organization. I have always been struck by the dedication of the ISRO staff and their identification with the goals of the space programme. The determination to see a project through to successful completion is not restricted to the senior management or only the scientific cadres. Technical, administrative and support staff willingly put in long hours, sometimes without monetary recompense. I believe such dedication and sense of involvement has played an important part in ISRO's success. This book was written primarily to tell the tale of what all these people together achieved in the field of launch vehicles. When I have pointed out what I perceive as shortcomings, it has been motivated solely by the desire that ISRO and its launch vehicle programme should continue to grow and prosper.

This book was possible only because many ISRO staffers, including those who are no longer with the organization, were willing to spend time with me, sharing their experience, explaining technical issues and patiently answering my never-ending doubts and queries. Both Prof. Satish Dhawan and Dr U.R. Rao, past chairmen of ISRO, readily granted me interviews. I am grateful to the present ISRO chairman, Dr K. Kasturirangan, for the encouragement and support I have received. I benefited greatly from discussions with a number of people, including Prof. E.V. Chitnis, Dr A.E. Muthunayagam, Dr Vasant Gowariker, Dr R. Aravamudan, Dr S.C. Gupta, Mr P.P. Kale, Dr R.M. Vasagam, the late Dr M.R. Kurup, the late Dr S. Srinivasan, Dr N. Vedachalam, Y.S. Rajan, Sreenivasa Setty, T.N. Seshan, P. Sudarsan, N. Sampath, G. Madhavan Nair, D. Narayana Moorthi, V. Sundararamaiah, Dr Rajaram Nagappa, Dr B.N. Suresh, Dr S. Vasantha, Dr V.N. Krishnamurthy, Dr M.C. Uttam, Dr K. Sitaram Sastri, S.K. Athithan, V. Sudhakar, P.R. Sadasiva, Dr M.K. Mukherjee, R.D. John, C.R. Sathya, Dr P.P. Sinha, R. Jeyamani, M.S.R. Dev, S. Ramakrishnan, K.L. Valliappan, Dr G. Viswanathan, Dr Anantharaman, Dr A.C. Bahl, Dr Natarajan and Dr Manoranjan Rao.

Dr P. Rama Rao, who was once director of the Defence Metallurgical Research Laboratory and is now the vice-chancellor of the University of Hyderabad, found time in his busy schedule for a couple of long discussions over telephone about the problems which had to be overcome in maraging steel development. Dr C.V. Sundaram, former director of the Indira Gandhi Centre for Atomic Research at Kalpakkam, too was good enough to give me the benefit of his years of experience in metallurgy.

It was as a reporter for *The Hindu* in Trivandrum that I first began to write about ISRO and its launch vehicle programme. But only after I became its science correspondent in 1991 did I seriously study the space programme and come to appreciate many of the technical intricacies involved. When I wanted to write this book, *The Hindu*'s editor, N. Ravi, and its executive editor, Malini Parthasarathy, generously permitted me to work on it without having to take leave. I am deeply grateful for their support as this book would have been impossible without it. But I must emphasize that the views expressed in this book are solely my own and should not be taken to be endorsed by *The Hindu* in any fashion.

Y.S. Rajan, former scientific secretary of ISRO and now scientific secretary to the principal scientific adviser to the Government of India, and Prof. S. Chandrasekhar, formerly with the ISRO headquarters staff and currently a faculty member at IIM Bangalore, were in large part responsible for pushing me into writing a book about ISRO. Chandra paid for this rash act by having to read through drafts of the chapters. His incisive comments and criticisms were of immeasurable value to me. To try to ensure that I had got my facts right and had not oversimplified issues, drafts of various chapters were shown to several senior ISRO scientists, some of whom were no longer with the organization. I am grateful to them for having found the time to go through the drafts carefully. Their comments and suggestions were invaluable. Needless to say, none of these people are responsible for any errors which remain in the book or for the views expressed.

S. Krishnamurthy, director of Publications & Public Relations at ISRO headquarters, with his customary cheerfulness and efficiency, coped with my endless requests for information, documents, photographs and permissions. The Central Documentation Division of the Vikram Sarabhai Space Centre made available many of the photographs which have been used in this book, especially those pertaining to the early days. Dr Manoranjan Rao let me photostat his entire collection of *Countdown* issues.

A. Ratnakar, the librarian at the Raman Research Institute, I.R.N. Goudar and his staff at the library of the National Aerospace Laboratories, as well as the staff of the libraries at ISRO headquarters and the ISRO Satellite Centre went out of their way to help me. The library at the Department of Atomic Energy headquarters in Mumbai provided photocopies of some of the early *Annual Reports* and articles from *Nuclear India* dealing with space.

In these days of the internet, I got help from people whom I had never met and came to know only through their web sites. My thanks, then, to Dr Jonathan McDowell of the Harvard Smithsonian Institute of Astrosphysics. Dr McDowell, who publishes an e-mail newsletter on space, *Jonathan's Space Report* (http://hea-www.harvard.edu/QEDT/jcm/space/space.html), sent me his paper on the Scout launch vehicle published in the *Journal of The British Interplanetary Society,* as well as a user manual about the Scout launcher prepared by the LTV Astronautics Division in early 1965. Both these documents helped me better understand the Scout vehicle on which India's first launch vehicle, the SLV-3, was modelled. Dr Jeff Foust, editor of the e-mail newsletter on space, *SpaceViews* (http://www.spaceviews.com/), sent me information about early communication satellites. I got in touch with Asif Siddiqi, who has written extensively on the Soviet/Russian space programme. He e-mailed me information about the history of the KVD-1 and the 11D56 engine which I have used. My thanks to him too.

Several colleagues went out of their way to help. Mahesh Vijapurkar, *The Hindu*'s deputy editor in Mumbai, rushed me photocopies of *Nuclear India* articles which I needed in a hurry. I am also grateful for the help I received from Dr R. Ramachandran, special correspondent for the *Frontline* magazine in Delhi, R. Krishnakumar, *Frontline*'s special correspondent in Trivandrum, and from A. Vinod, *The Hindu*'s special correspondent for Sports in Trivandrum.

At Penguin, Krishan Chopra and Aradhana Bisht had to cope with a difficult manuscript and probably an equally difficult author. It was, however, a task they carried out efficiently and with much tact. My heartfelt thanks to both of them.

Credits for Photographs and Diagrams

The Department of Space/Indian Space Research Organization supplied all but one of the photographs used in this book. The photograph of the early Indian group during their training at NASA's Wallops Island facility which appears on page 15 was provided by Dr R. Aravamudan.

Many of the diagrams used in this book are based on ones published by DOS/ISRO. These include the diagrams of the RH-125 to SLV-3, the SLV-3, SLV-3 Trajectory, ASLV Configuration, Flight Sequence of ASLV, An early PSLV Configuration, Current PSLV Configuration, PSLV-C2 Flight Profile, SLV-3 to PSLV, and PSLV to GSLV.

Most of these diagrams had to be modified, principally to simplify them. In addition, some fresh diagrams had to be drawn. Both these tasks were ably carried out by M/s Professional and M/s Taracorp.

CHAPTER

1

The Sarabhai Vision

It is not difficult to picture the scene. In 1962, probably in the month of November, amid a thicket of coconut trees and the ceaseless roar of waves, two men stood in animated discussion in a fishing village on the outskirts of Trivandrum (now called Thiruvananthapuram). There is, in fact, a photograph of them together at that spot. Although there were others who had accompanied them, it was quite obvious that these two men were the leaders, the ones who made the big decisions.

One of them had an infectious, boyish enthusiasm which made him seem younger than his forty-odd years. This was Vikram Sarabhai. The older man with him was none other than Homi Bhabha, a patrician figure who had been responsible for starting India's nuclear programme.

Almost certainly, curious fishermen, their wives and children would have watched this group's every move, keeping a respectful distance but speculating aloud about

Vikram Sarabhai and Homi Bhabha

the purpose of their visit. They could never have guessed that in a year's time, their village would become the Thumba Equatorial Rocket Launching Station (TERLS). Even this, as it turned out, was only the beginning. Sarabhai's dreams would lead to the creation of a full-fledged space programme. As a result, India is today able to design and build its own remote sensing and communication satellites. Moreover, the Indian Remote Sensing (IRS) satellites are being launched by its own Polar Satellite Launch Vehicle (PSLV).

It was Sarabhai's vision which gave shape to the space programme. It was he who enunciated the need for such a programme and set its goals for decades to come. These have continued to be the principles which rule the programme. So in order to understand the space programme and its purpose, it is necessary to know more about Sarabhai and his thinking.

The man behind the name

Vikram Ambalal Sarabhai was born in Ahmedabad on 12 August 1919 to a wealthy family of industrialists. His early education was in a family school run by his mother. The school embodied Maria Montessori's dictum that a teacher's function was to inculcate the love and pursuit of knowledge. One who knew the family wrote about the 'simple life of refinement and culture' which the Sarabhais led. Despite the family's wealth, luxury and ostentation was disapproved of. Vikram grew up learning the importance of hard work, a value he exemplified throughout his life.

He grew up to be, as one of his teachers recalled, 'a handsome young boy with a lovable personality, pleasant manner, courteous behaviour and a sharp intelligence'. These traits would always mark him out and endear him to people.

He joined the Gujarat College in Ahmedabad but left before graduating to join Cambridge University. There, in 1940, he took his Tripos in Natural Sciences. When World War II broke out, his father insisted that Vikram return to India. He then came to Bangalore and joined the Indian Institute of Science, where C.V. Raman was director.

S. Ramaseshan, a well-known scientist in his own right, remembers meeting Vikram at the 1941 annual meeting of the Indian Academy of Sciences. At the time, however, Ramaseshan was in his first year of B.Sc.(Hons). More than five decades later, he gave a vivid description of that first meeting with Sarabhai:

> Then came Vikram Sarabhai. He was scarcely 21. He looked like a boy just out of school. He had the beautiful skin of a baby—it actually glowed. It could only be described by the untranslatable Tamil expression 'Paal Vazhiarathu'. A baby's skin as though milk is still flowing over it. Raman introducing him said, 'Young Vikram Sarabhai has been brought up with a silver spoon in his mouth. He has started to do original experiments and is presenting his first paper to a scientific audience. I have great faith in him—that he will contribute much to India

Vikram Sarabhai

and to the growth of science in our country. We are sure to hear his name again and again in many contexts.'

Although Sarabhai returned to Cambridge after the war, the Bangalore years left an indelible mark on him. For one, there was Raman himself. Raman believed that in order to make a mark in science, Indians needed to take advantage of the unique opportunities available within the country. He himself had chosen to study the scattering of light because of the plentiful sunshine. Later when Sarabhai established a sounding rocket launching station and identified applications for India's space programme, his reasoning reflected this line of thought.

In Bangalore, Sarabhai met two people who would be closely associated with his life and work. One was Homi Bhabha. Bhabha had joined the faculty of the Indian Institute of Science shortly before Sarabhai arrived. Sarabhai became part of Bhabha's small group working on cosmic rays, the high energy particles that continually bombard the Earth. He continued working on cosmic rays after returning to Cambridge and in 1947 received a doctorate for his thesis on 'Cosmic Ray Investigations in Tropical Latitudes'.

Homi Bhabha

Bhabha and Sarabhai had much in common. Both came from wealthy families and shared a serious interest in science as well as the desire to ensure its growth in India. Their close ties continued long after they had both left Bangalore and the Indian Institute of Science. Almost two decades later, Sarabhai would create the space programme under the protective umbrella of Bhabha's Department of Atomic Energy. When Bhabha died in a plane crash in 1966, Sarabhai inherited his mantle as head of the atomic energy programme.

But professional interests were not all they shared. Ramaseshan recounts that 'Bhabha and he [Vikram] were seen almost every day at West End [still one of Bangalore's most expensive hotels] and the clubs surrounded by beautiful women, much to the envy of all at the Indian Institute of Science'.

It was in Bangalore that Sarabhai met Mrinalini Swaminathan, whom he married. Both of them had strong professional interests to which they were deeply committed, Sarabhai to science and Mrinalini to dance. As their daughter, Mallika, narrates, 'On their first evening out, Papa clarified to Amma, "I am committed to science and never want to marry." To his surprise, Amma looked relieved and said, "Thank God. I am so tired of being proposed to.

I too want only to be a dancer. Now we can be friends." Six months later they were married.'

Soon after getting his doctorate in 1947, Vikram Sarabhai returned to India. He was asked by the Ahmedabad Mill Owners' Association to establish the Ahmedabad Textile Industry's Research Association (ATIRA). He was just twenty-eight, and had no experience in textiles or training in textile technology. Yet, this young man had very clear ideas about the path to follow and convinced others that this was a sensible course. He did not hire people with a textile technology background or those with experience. Instead, he appointed young people with a good scientific training.

Sarabhai would later apply this same principle when he started the space programme. He was less interested in experience and looked primarily for good scientific training, plenty of enthusiasm and the ability to work hard. He wasn't afraid of failure and would encourage those willing to take calculated risks. This philosophy seems to have paid off and ATIRA prospered. It was able to show results in just three years and began to play an important role in a number of issues vital to the industry and the nation, says Kamla Chowdhry, a long-time associate of Sarabhai. ATIRA's success led to similar textile research institutions being set up in Bombay (now renamed Mumbai) and Coimbatore.

One would have thought that Sarabhai had quite enough on his hands with ATIRA. But 1947 was also the year he started the Physical Research Laboratory (PRL) at Ahmedabad. In the initial years at least, the small research group at PRL worked on Sarabhai's field of interest, cosmic rays. In 1960-61, PRL became an autonomous institution under the Union government's Department of Atomic

Energy. Till a separate Department of Space was created soon after Sarabhai's death, the Department of Atomic Energy routed funds for the space programme through PRL. It was PRL which provided the administrative support needed for the programme in those early years.

This extraordinary man, builder par excellence of institutions, wasn't through yet. He took over management of Sarabhai Chemicals in 1950, established Suhrid Geigy Limited in 1955, assumed the management of Swastik Oil Mills Limited, founded the Ahmedabad Management Association in 1957, and set up Sarabhai Merck Limited in 1958. He also took over Standard Pharmaceuticals in Calcutta and found time and energy to establish both the Sarabhai Research Centre at Baroda and the Operations Research Group (ORG) in 1960. The following year, he set up three more companies. He was also the prime mover behind establishing the Indian Institute of Management (IIM) at Ahmedabad, which remains one of the country's most prestigious centres for management training.

Sarabhai could not only wear multiple hats, he could also wear them well. He used computers to analyse the marketing methods of Swastik Oil Mills. The findings made it possible to strengthen the company's marketing system. The ORG continues to carry out market research studies for private and public companies. In the field of pharmaceuticals, he took the lead in ensuring quality controls and indigenization of raw materials used in drug manufacture. He worked twenty hours a day and considered sleep a luxury. He always found time for all his various duties.

This, then, was the man who would found the Indian space programme.

Starting a sounding rocket station

The Indian space programme began with the setting up of a station to launch sounding rockets. Sounding rockets get their name because they 'sound' the atmosphere — in other words, measure various parameters — during their flight. Unlike a launch vehicle which has to put a satellite into orbit, a sounding rocket has only to carry instruments straight up. When its propellants are exhausted, the rocket falls back to earth.

The Space Age had dawned in a dramatic fashion when the then Soviet Union launched the world's first satellite, the Sputnik, on 4 October 1957. By the early Sixties, there were quite a few people at PRL who had just returned after studying in the United States and who were aware of research possibilities offered by space technology. 'We were doing space science from the ground, and so interest in space and making *in situ* measurements was a natural extension,' says E.V. Chitnis, who was a close associate of Sarabhai at PRL and later in the space programme.

Years later, when he delivered the Bhabha Memorial Lecture to the Indian Rocket Society in December 1969, Sarabhai explained the importance of collecting information about the upper atmosphere:

> In the field of study of our environment, one of the most important areas is the one which ranges from about 40 kilometres to about 200 kilometres, an area which is not accessible to balloons and which is below the operational altitudes at which satellites can operate. It is this area which is most importantly covered with sounding rockets. As one knows very well, sounding rockets can only perform if ground facilities and basic back-up are available. The study of this region in the equatorial areas is one of the major gaps in the study of

> our environment today. And so, as far as India is concerned, with the facilities that have grown up, we have fantastic opportunities in the years to come to understand many complex phenomena involving the interaction of the ionosphere with the geomagnetic field, problems of the neutral and the ionised atmosphere and the interaction of these two. These subjects are of importance not only for the understanding of radio propagation, but also from the point of view of meteorology and basic problems of energy and momentum transport into the lower atmosphere where climate is made.

The 'equatorial electrojet' was one of the phenomena to be studied with sounding rockets. The equatorial electrojet is a stream of electric current flowing in a narrow band of about three degrees on either side of the magnetic equator at a height of some 100 km. Early studies had been made of the electrojet with ship-based rocket launchings in the Pacific. A sounding rocket facility on the magnetic equator would allow detailed analysis of the electrojet's vertical structure and fluctuations.

In August 1961, the Union government placed space research and peaceful uses of outer space under the jurisdiction of the Department of Atomic Energy (DAE). In February 1962, the DAE created the Indian National Committee for Space Research (INCOSPAR) under the chairmanship of Vikram Sarabhai to oversee all aspects of space research in the country.

The UN Committee on the Peaceful Uses of Outer Space had passed a resolution recommending the creation and use of sounding rocket launching facilities, especially in the equatorial region and the southern hemisphere, under United Nations sponsorship. In 1962, the Committee on Space Research (COSPAR), an international scientific body, had pointed out the importance of the equatorial region for meteorology and aeronomy. Noting the gaps in the world coverage of sounding rocket sites, it urged the establishment of a sounding rocket launch station on the magnetic equator under UN sponsorship.

These pleas dovetailed neatly with Sarabhai's own plans. Agreements were quickly signed with the National Aeronautics and Space Administration (NASA) of the United States for training in assembling and launching imported sounding rockets and help in establishing the launch station. The United States also supplied many of the sounding rockets which were launched in the initial years. Soon afterwards, an agreement was signed with the French space agency, Centre National d'Etudes Spatiales (CNES). The French supplied a radar and also sounding rockets. Later, British and Russian sounding rockets were also launched. These sounding rockets carried aloft a variety of instruments for making different kinds of measurements.

By mid-1962, Chitnis, who had become member-secretary of INCOSPAR, was despatched to locate a site on the west coast suitable for setting up a sounding rocket launch station. The magnetic equator passed quite close to the town of Quilon (now called Kollam) in Kerala, just 60 km north of the state capital, Trivandrum. He made the trip down from Ahmedabad by plane. From the Dakota aircraft, Chitnis could see the coast of southern Kerala where the launch site had to be found. He was received at the airport by the district collector of Trivandrum and promptly set out to look at possible sites. One of the earliest he visited, if not the first, was the fishing village of Thumba, not far from the airport. During that and subsequent trips the suitability of various sites along the coastal belt from Trivandrum up to Alleppey (now called Alappuzha) were examined. One of those trips was with an American team from NASA's Wallops Island sounding rocket facility. The exact position of the magnetic equator was determined during this visit, says Chitnis.

According to Chitnis, Bhabha and Sarabhai flew down together to Kerala to finalize the site around November that year. After taking off from Cochin (now Kochi) en route to Trivandrum, the pilot was asked to fly low so that Bhabha and Sarabhai could get a bird's-eye view of the coastal areas. Bhabha sat in the co-pilot's seat with Sarabhai standing behind him. The other passengers may well have wondered, possibly with some alarm, why the aircraft was flying so low, just skimming the treetops.

A fleet of cars was waiting at the Trivandrum airport and they straightaway left for Quilon to inspect the sites there. One of the sites, some 25 km from Quilon, had the unfortunate name 'Vellana thuruthu'. In Malayalam, 'Vellana' means 'white elephant' and 'thuruthu' means 'island'. As Sarabhai later remarked, 'We steered clear of it, for fear of it becoming a national joke.'

Next day, Bhabha and Sarabhai returned to Trivandrum, and stayed at the Raj Bhavan as guests of the governor, V.V. Giri. They visited both Thumba and the nearby Veli Hill which overlooked it. In Kerala's hot and humid conditions, Bhabha was tired and covered in sweat by the time he reached the top of the hill. He apparently then decreed that buildings there had to have air-conditioning. When those buildings were built some years later, they were in fact air-conditioned.

In the end, the decision was in favour of Thumba. Although it was not directly below the magnetic equator, sounding rockets launched from Thumba would be able to probe phenomena associated with it, such as the equatorial electrojet. In an article published a year later, Sarabhai said that Thumba had been preferred to the Quilon area because fewer people needed to be rehabilitated and there was less fishing activity. The latter was also an important consideration because spent stages from the sounding rockets would fall into the sea. According to Chitnis, Thumba had the added advantage of an airport close by so that people and material could be flown in when needed.

The decision to establish the sounding rocket launching station at Thumba was made known to Parliament on 21 January 1963, when Lakshmi N. Menon, minister of state for External Affairs and herself from Kerala, answered a question on behalf of the Prime Minister.

The district collector of Trivandrum was asked to acquire the land and hand it over in just a hundred days. Land acquisition by the government is problematic in most parts of India. Such problems are greater in Kerala which has a higher population density than other regions, particularly so along the beaches where the fishermen live. They lead a hard life, with more than its share of uncertainity, and tempers are apt to flare quickly. A wrong move in such an emotive issue as removing people from the land they are living on could easily have created an explosive situation.

The credit for carrying out the land acquisition with tact and sensitivity goes to the collector of Trivandrum, K. Madhavan Nair, and the bishop of Trivandrum, the Right Rev. Dr Peter Bernard Pereira. Since the fishermen were Christians, the words of the bishop carried much weight. Instead of the thatched huts they had occupied, brick and mortar houses were built for them at a site nearby. The Department of Atomic Energy and the bishop contributed to building these houses. As people were moved to these houses, their land was handed over to the Central government. The collector and his deputy visited the site every day to check on the progress and to hear any complaints. As Madhavan Nair records, the bishop spared no pains in supervising the rehabilitation of the displaced people. In this manner, one square mile was acquired on the beach at Thumba, along with a hundred acres of land on the Veli Hill.

Not only did the Catholic church help in securing the land at Thumba, they also agreed to vacate the St Mary Magdalene's Church which fell within the area to be acquired. This church and its parsonage became the first laboratories and offices of the space programme. The church has been preserved and is currently maintained as a space museum.

R.D. John, who took charge of civil works at Thumba and later became the Department of Space's chief engineer, heading its Civil Engineering Division, recalls flying down to Trivandrum with Chitnis in January 1963. When he asked for a copy of the project report, Chitnis gave him a single sheet which listed various works totalling just Rs 15 lakh. John thought it a small job which could be completed in six months and wondered what he could

St Mary Magdalene's Church

find to do after that. He need not have worried. As he subsequently remarked, the space programme blossomed and construction became an unending affair.

The first launch from Thumba

In the meantime, Sarabhai had recruited a small group of young men and sent them to NASA for training at the Goddard Space Flight Centre and at the Wallops Island facility used for sounding rocket launches. The training was only in assembling imported sounding rockets and their scientific payloads, procedures for the safe launch of these rockets, tracking the flight of the rockets, receiving data radioed down during flight and collecting other scientific information required. On their return, these people would create and operate what came to be called the Thumba Equatorial Rocket Launching Station (TERLS).

R. Aravamudan, a young man of twenty-three working at the Atomic Energy's research centre in Trombay (which later became the Bhabha Atomic Research Centre), was glad to get the chance to leave Bombay. Aravamudan later became director of the Sriharikota launch centre and subsequently of the ISRO Satellite Centre. Pramod Kale, even though he was still only a student of Sarabhai's, was also sent. Kale afterwards became the director of the Space Applications Centre in Ahmedabad and then of the Vikram Sarabhai Space Centre. A.S. Prakasa Rao and B. Ramakrishna Rao, both from PRL, were part of the same group. H.G.S. Murthy, who would head TERLS on his return, went a couple of months later. Abdul Kalam, who moved over from Aeronautical Development Establishment (ADE), a defence laboratory in Bangalore,

L-R: R. Aravamudan, Abdul Kalam, H.G.S. Murthy, B. Ramakrishna Rao, D. Easwardas
(courtesy R. Aravamudan)

and D. Easwardas, who came from the atomic energy programme, joined the others shortly afterwards.

By mid-1963, they were back in India and preparing for the first launch of a sounding rocket from Indian soil. Aravamudan wrote this vivid account of those days:

> Coming straight from NASA, both Trivandrum and Thumba proved to be a shock. Our facilities at Thumba consisted of one launcher, a church and some old fishermen's dwellings — a very far cry indeed from the luxury in terms of equipment and facilities which we had got used to at Washington DC and Wallops Island. At Trivandrum, too, it was difficult to get convenient accommodation, and food was a problem, unused as we were to Kerala cooking. Even getting to the office and back was very tough, since we had to rely on KSRTC buses [the Kerala State Road Transport Corporation's city

bus services]. Every morning, we would walk up to the railway station, eat breakfast at the canteen, pick up some packed lunch and take a bus. We would pay 90 paise and the bus would take a long winding route through Kazhakuttam and Pallithura. It would take us almost an hour to reach TERLS. At Thumba, we sat in the church building, which we shared with generations of pigeons. All those facilities which Sarabhai had talked of were still very much a dream.

On 21 November 1963 a Nike-Apache rocket supplied by NASA was assembled in the church building, along with a sodium vapour experiment provided by CNES of France. American and French technicians were present to help the Indians on this occasion. The rocket was then moved by truck to the launch pad. As it was being hoisted onto the launcher, the hydraulic system of the crane sprang a leak. There was no time for repairs. Fortunately, it was possible to manually lift the heavy rocket and place it on the launcher.

Nike-Apache

That wasn't the end of the problems, according to Kale. The remote system to raise the launcher to the correct angle for launch did not work properly and a man had to be sent out to operate the controls on the launcher itself. Five minutes before launch, an alarm was sounded to clear the area around the launch pad. Two minutes later, Kale looked out of the blockhouse to find the man still adjusting the launcher. 'I ran out and got him inside,' says Kale.

At 6.25 p.m., the Nike-Apache rocket streaked away into the gathering dusk. Some minutes later, a sodium vapour cloud emerged high above, tinged orange by the rays from the setting sun and visible from places far away. The movement of the cloud indicated the winds prevailing at that height. 'An orange-coloured cloud was visible to the naked eye at places up to 250 kilometres away,' wrote Sarabhai subsequently. He added:

> It was planned to photograph the cloud from special cameras at places as far separated as Kanyakumari, Palayamkottai, Kodaikanal and Kottayam. However, the sky was completely overcast at Kodaikanal and partially so at Kottayam at the time of the firing. Records have been obtained from the other two stations. Putting together information from these stations where the cloud was photographed against the background of stars, it will be possible to gain fresh insight into the complicated problems connected with the electrojet and high altitude aeronomy in the Indian Ocean area.

It marked the beginning of the Indian space programme.

The need for a space programme

A year later, an agreement had been signed with a French company to produce their Centaure sounding rockets under licence in India. The following year, the government gave sanction for starting the Space Science and Technology Centre (SSTC). 'The objectives of the Centre are to design, develop and construct rocket and satellite payloads and instrumentation and to promote research in the space sciences and technology,' stated the *1966-67 Annual Report* of the Department of Atomic Energy. The SSTC would initially gain experience by building payloads for sounding rockets and developing indigenous sounding rockets. But Sarabhai clearly envisaged a time when India would be able to build satellites and put them into space on its own launch vehicles.

Chitnis and Kale, who worked closely with Sarabhai at the time, confirm that Sarabhai had a full-fledged space programme in mind from the very beginning. Kalam, in his autobiography, *Wings of Fire*, recalls that 'after the successful launch of Nike-Apache, he [Sarabhai] chose to share with us his dream of an Indian Satellite Launch Vehicle'.

It is not likely that people like Sarabhai — or Bhabha would have been content simply to fire imported sounding rockets or licence-produce them in India. The fact that land on Veli Hill was acquired at the same time as Thumba indicates that much more than a launch facility for sounding rockets was contemplated right from the beginning.

Just two months after the Thumba site had been selected, and several months before even an imported sounding rocket could be launched from Indian soil, Bhabha inaugurated an international space physics seminar at Ahmedabad on 28 January 1963. In his speech, he argued that a beginning had to be made in the field of space research so that the country did not fall behind others in practical technology. 'If we do not do so now, we will have to depend later on buying know-how from other countries at much greater cost.' Bhabha referred in particular to the possibilities opened up by communication satellites.

The Sixties were hardly a propitious time in India for so ambitious a venture. November 1962, when Sarabhai and Bhabha flew down to Kerala to select the site for the sounding rocket launch station, was also the time when India fought and lost a war with China. India was still desperately poor and dependent on food grains imports. Many within the country and more so abroad questioned the need for a country so impoverished to embark on a nuclear and later a space programme. After all, these programmes had to be supported by financial resources which could be usefully deployed in other ways. But Bhabha and Sarabhai, like Pandit Nehru, firmly believed that a strong indigenous base in science and technology was essential for India to pull itself out of the state of underdevelopment and provide self-sustaining growth.

When Prime Minister Indira Gandhi formally dedicated the Thumba Equatorial Rocket Launching Station to the United Nations on 2 February 1968, Sarabhai, in his speech, clarified the purpose of the Indian space effort:

> There are some who question the relevance of space activities in a developing nation. To us, there is no ambiguity of purpose. We do not have the fantasy of competing with the economically advanced nations in the explorations of the moon or the planets or manned space flight. But we are convinced that if we are to play a meaningful role nationally, and in the community of nations, we must be second to none in the application of advanced technologies to the problems of man and society which we find in our country.

Should a poor, developing country like India have a space programme? Didn't it have much more important priorities like providing its people with adequate food, clothing, housing and education? Sarabhai, however, argued that development of a nation was closely linked with a

Prime Minister Indira Gandhi dedicates the TERLS to the UN

good understanding of science and the sound application of it in the form of technology.

A crucial part of the thinking of Sarabhai, Bhabha and Pandit Nehru was that a strong science and technology base would help India leapfrog some of the stages of development which the Western nations had passed through. According to Kamla Chowdhry, Sarabhai would often say, 'I have a dream, a fantasy maybe, that we can leapfrog our way to development.'

In *Atomic Energy and Space Research: A Profile for the Decade 1970-80* published in 1970, Sarabhai, who had assumed leadership of the atomic energy programme as well after Bhabha's death in 1966, wrote:

> The progress of science and technology is transforming society in peace and in war. The release of energy of the atom and the conquest of outer space are two most significant landmarks in this progress. Largely due to the consistent national support which the programmes of the

> Atomic Energy Commission has received since Independence, India is amongst the nations of the world advanced in atomic energy, and is striving for a similar position in space technology and research. There are those who preach that developing nations must proceed step by step following the same process by which the advanced nations themselves progressed. One is often told that such and such a thing is too sophisticated to be applied. This approach disregards what should perhaps be obvious that when a problem is great, one requires the most effective means available to deal with it.
>
> The seeming disadvantage of a developing nation such as India, which has only a limited existing technological infrastructure to build on, can be an asset rather than a liability. I suggest that it is necessary for us to develop competence in all advanced technologies useful for our development and for defence, and to deploy them for the solution of our own particular problems, not for prestige, but based on sound technical and economic evaluation as well as political decision-making for a commitment of real resources.

In the same piece, Sarabhai very tellingly observed: 'There might be many opinions concerning what would be an advantageous course to follow in the short run, but ten to twenty years from now, when the population of India would be somewhere between 750 and 1,000 million, it can hardly be controversial that we would need a very strong base of science and technology, of industry and agriculture, not only for our economic well-being but for our national integration and for ensuring our security in the world.'

Sarabhai was a person with a wide circle of friends in the international scientific community and kept abreast of developments in space technology. PRL had, for instance, a facility to receive data directly from NASA's early scientific satellites. His genius was in recognizing very early the practical benefits of space applications for a developing country and one as large as India.

He identified three specific applications: remote sensing, communications and meteorology. These three have remained the *raison d'etre* of the Indian space programme. Today such applications are routine and taken for granted the world over. But in the Sixties, the applications and the technology which made them possible were still new, even in the United States.

Telstar 1, the world's first true communications satellite, was launched only in 1962. With amazing foresight, Sarabhai saw the difference that direct TV broadcasts from satellites could make, a technology which would come into its own only in the Eighties and Nineties. Bhabha's speech of January 1963, which reflects Sarabhai's thinking, predicted that satellite communication 'may well supersede our present means of inter-continental communication' and that 'it could also, for the first time, make inter-continental or world-wide television possible'.

In a paper presented at a conference in 1969, Sarabhai, however, pointed out the enormous impact of reaching television to about 80 per cent of India's population. Television was the ideal medium for conveying news and important developmental information to the people, a large proportion of whom were illiterate. Direct broadcast of TV by satellite provided, for the first time, an advanced technology which could reach information to villages all over the country just as easily as to urban areas.

As early as 1967, a joint study had been conducted with NASA to examine alternatives for television coverage of India. The study concluded that a hybrid system — which India now has in place — involving the use of direct broadcast by satellite as well as retransmission through conventional terrestrial TV stations would be the most cost-effective solution. Two years later, in September 1969, an agreement was signed with NASA for loan of its ATS-6 satellite to experimentally broadcast educational TV programmes to villages in India. The Satellite Instructional Television Experiment (SITE) exercise, which took place in 1975-76, was probably the first time that direct reception of TV programmes beamed via satellite was attempted anywhere in the world on such a scale. Some 2,400 villages in various states were equipped with direct reception community TV sets for the experiment.

In the second half of 1970, another joint study was conducted, this time with the Massachusetts Institute of Technology, on the design for the Indian National Satellite (Insat). These were Sarabhai's initiatives and they laid the foundations for the Insat satellites. The Insat-1 satellites were made in the United States. The first of the Insat-2s, designed and built in India, flew in 1992.

In the field of remote sensing, Landsat 1, the world's first commercial remote sensing satellite, was launched by the United States only in July 1972. In a talk delivered in December 1970, Sarabhai envisaged precisely the sort of applications in agriculture, forestry, oceanography, geology and mineral prospecting, and cartography which the Indian space programme would later take up and foster. It is only in recent years that these applications have taken root among government departments and other user agencies. The Indian Remote Sensing (IRS) satellites today form the largest constellation of remote sensing satellites.

The earliest meteorological satellites, the Tiros series, were launched from 1960 onwards. But these were orbiting satellites. The *Atomic Energy and Space Research Profile for*

1970-80, however, speaks of using satellites in geostationary orbit (where it would keep pace with the Earth's rotation and therefore appear stationary from the ground) for communications as well as meteorology. A meteorological camera in geostationary orbit can continuously watch over rapidly evolving phenomena, such as cyclones, something that orbiting satellites cannot do.

Sarabhai continually emphasized the importance of self-reliance and self-confidence. He has been quoted as saying that 'we do not wish to acquire black boxes from abroad but to grow a national capability'. He wanted India to build satellites which would deliver the applications he saw as being of vital importance to the country as well as the launch vehicles to put those satellites into orbit. This would remain the ruling tenet of the space programme.

The early days of SSTC

The Space Science and Technology Centre (SSTC) was intended to be the technological powerhouse of the space programme. For it, Sarabhai recruited several young Indians who were abroad, almost all in the United States. All of them had doctorates and many were already employed there. A.E. Muthunayagam, who would take charge of liquid propulsion activities, was working on a NASA project in the United States. S.C. Gupta, who would build up the group which would make control and navigation systems for launch vehicles, was working for a company which was one of the first to use digital computers for automatic process control in the industry. M.K. Mukherjee was teaching at a university and handled materials development after joining the space programme. Others who joined at the same time included Y.J. Rao who later assumed responsibility for building up the facilities at Sriharikota, D.S. Rane, who established the computer facilities and programming capability needed, and M.C. Mathur. Vasant Gowariker, a chemical engineer, was recruited by Sarabhai from Britain.

They had responded to an advertisement in a publication of the Indian embassy in Washington DC announcing job opportunities in the space programme. One cannot help wondering if any Indian living today in the United States would apply if a similar advertisement appeared now! Sarabhai personally interviewed and selected each of them.

But most people were recruited from India itself. R.M. Vasagam, who later played an important part in both launch vehicle and satellite development, was one of those who joined early. Some, like M.R. Kurup and M.C. Uttam, both of whom were closely involved in the successful production under licence of Centaure sounding rockets in India, came from the atomic energy programme. Others, including G. Madhavan Nair, who became the project director for the Polar Satellite Launch Vehicle and is currently director of the Vikram Sarabhai Space Centre, were recruited from Atomic Energy's Training School. As the scope of the programme expanded and the need for personnel grew, people joined from a wide variety of institutions.

For those who came after living and working in the West, the transition to Trivandrum and Thumba would have been dramatic, if not traumatic. It wasn't much better for those recruited within the country. Although it is the capital of Kerala, Trivandrum was then, and to a large extent still is, little more than a town. Almost all who were recruited abroad were married and so could set up home. Most of those recruited within India were younger and bachelors. In a staunchly non-vegetarian state, the vegetarians — who were quite a few — had a particularly difficult time. The restaurant at the railway station was probably the only place which served decent vegetarian food. To this day, choice of restaurants, let alone cuisine, is quite limited in Trivandrum. Many of the bachelors lived in a lodge near the railway station. Trivandrum had also little to offer by way of entertainment or relaxation.

The new buildings for SSTC on Veli Hill were not yet ready. So people were accommodated in the church buildings and some adjacent sheds. There were few vehicles owned by the space programme and people went about in Thumba on bicycles. Bicycles were even used, on occasion, to move payloads, which would later be launched on a sounding rocket, from one place to another.

One of the early issues of *Countdown*, house journal of the Vikram Sarabhai Space Centre, had this account of 'The Bicycle Era':

> Those were the days (1963-1965) when the progress of the Thumba station was controlled by a single vehicle. A green-coloured Standard van was the workhorse that ferried mail, people, coffee, stones and chemicals. There was no point within the Trivandrum city limits that was not touched by this remarkable vehicle — provided there was a road to go.
>
> All other movements were the monopoly of the bicycle, specially on the campus. The Heads of Divisions would all have

The nose cone of a sounding rocket being transported

> their bicycles reserved for 'inspection rounds' during fixed time slots. They would hop on it and pedal off to distant places in the hot sun; mercifully, the roads were level, for the Veli complex was yet to emerge. At other times, the rest of the staff would put the bicycle to all earthly uses that even the manufacturers could hardly conceive.

Although living and working conditions were far from ideal, very few people left the space programme in those early days. Why did so many talented people opt to stay on? Camaraderie was certainly one reason. Even today, despite all the differences and disputes, strong bonds tie those who were there during that period. They would probably agree with Aravamudan's observation: 'Tough as those days were, there was a warmth about them. A family atmosphere which has, sadly, gone forever.'

Even more important was the personality of Sarabhai himself. Sarabhai was able to give those he recruited and led the feeling that they were partners in a great venture which would benefit the country. Unassuming, dressed often in kurta-pyjamas, Sarabhai was an extremely approachable person. His daughter, Mallika, speaks about Sarabhai's ability in making 'every single person he dealt with feel that he or she was the most important person for him; and for that minute they were just that'. P.D. Bhavsar, who worked closely with Sarabhai at PRL and then in the space programme, wrote that even a peon or a driver could meet Sarabhai without fear or being made to feel inferior. To Sarabhai, all humans were equal and he strove to respect their dignity.

If the long-term goal was to build satellites as well as the launch vehicles which could put them in orbit, a beginning had to be made in developing the team work and basic technological competence needed by indigenously building payloads and sounding rockets. According to Chitnis, a favourite phrase of Sarabhai's was 'this is the thin end of the wedge'. In other words, get started in a small way and then expand. So people were set to work on different aspects of technology. People and groups could find themselves allotted several tasks. No area was the exclusive preserve of any one person or group. Indeed, often two or more people could be working on different approaches to the same problem. It was probably Sarabhai's way of encouraging the spirit of competition.

A year after SSTC was started, Bhabha died and Sarabhai took his place as head of the nuclear programme. Despite his new duties, Sarabhai still continued to visit Trivandrum regularly. He remained the operational head of the space programme. Proposals for any new projects or lines of work were made to him directly and he decided, often then and there, whether these should be pursued or not. As Kalam narrates, 'the very news of his coming to Thumba would electrify people'. All laboratories, workshops, design offices would hum with unceasing activity. People would work virtually around the clock because of their enthusiasm to show Prof. Sarabhai something new.

But there was a less rosy side too. After Sarabhai left, there would be confusion, recriminations and people would not talk to one another. H.G.S. Murthy, the head of TERLS, was also deputy-director of SSTC. But since he was not an R&D man, he was apparently held in scant respect and SSTC was, in Sarabhai's absence, to be ruled by the Technical Coordination Committee and later the Technical Coordination and Finance Committee (TCFC). I was told by several people that the members of the TCFC — Gowariker, Gupta, Mukherjee, Muthunayagam, and Y.J. Rao — were nicknamed the 'Pancha Pandavas'. 'Those fellows would go on fighting with each other,' remarked one observer.

Since the power to make important decisions was only with Sarabhai, such bickering did not get out of hand. With his frequent visits, Sarabhai could make sure that no one was suppressed. He would listen to everyone, encouraging them to do more. Although there is a distinct air of confusion about that period, with various people doing all sorts of disparate things, slowly things coalesced. Groups which could build sounding rockets within the country rapidly emerged. The indigenous development of sounding rockets is taken up in more detail in the next chapter.

These were just the first step towards launch vehicles. It would be a long and difficult journey. Only in October 1994, with the first successful launch of the Polar Satellite Launch Vehicle, did India achieve Sarabhai's dream of building and launching application satellites.

CHAPTER

2

First Steps in Rocketry

HAVING ESTABLISHED THE Thumba Equatorial Rocket Launching Station and gained some first-hand experience in launching foreign sounding rockets, building such rockets and payloads for them within the country was a necessary first step before attempting even the simplest of launch vehicles.

As the *Atomic Energy and Space Research: A Profile for the Decade 1970-80,* pointed out:

> The development of systems as complex as a satellite launch vehicle and a satellite needs understanding in depth and complete mastery of the technology of each subsystem which is involved. The thrust of the programme at the Space Science and Technology Centre during the past three years has been to grow this capability through a number of individual projects, each by itself modest in character, but progressively involving increasing technological complexity and sophistication.

The majority of young men whom Sarabhai had recruited for the fledgling space programme had no background in rocketry. A sounding rocket programme would give them hands-on experience and create basic competence in some of the important technologies. Sarabhai recognized that much more than just technical skills were needed for a space programme. In his foreword to the profile quoted above Sarabhai writes:

> There is a totality about modernisation, and in order to gain confidence, we must experiment within our resources even at the risk of failure. We have to rise from an in-built culture within which a major departure from an existing well-proven system and anything which is innovative in character is automatically regarded with suspicion.

The sounding rocket programme would give India's aspiring space scientists and engineers experience in conceiving projects and executing them with multi-disciplinary teams. In a paper published by the American Astronautical Society in 1966, Sarabhai pointed out that 'when a nation succeeds in setting up a scientific programme with sounding rockets, it develops the nucleus of a new culture where a large group of persons in diverse activities learns to work together for the accomplishment of a single objective'.

Moreover, indigenous sounding rockets would also probably be cheaper than imported ones. The United States, France, Britain and the then Soviet Union had provided India with sounding rockets for research programmes in which they were involved. But with their own rockets available, Indian scientists would be able to pursue their research without having to always tag along with an international effort.

About a year after the first sounding rocket was launched from Thumba, during 1964-65, an agreement was signed with Sud Aviation of France to build their two-stage Centaure rocket under licence in India. But Sarabhai was not going to be content with licensed production. He encouraged his people to develop sounding rockets on their own and this directly led to the Rohini sounding rocket programme.

In the early days, facilities to make even simple rockets did not exist at Thumba. Undaunted, the enthusiastic young men set to work, making do with whatever was available and improvising as they went along. They experimented with various solid propellant combinations. These were tested by stuffing them in small metal tubes, capped at one end and fitted with a simple igniter charge and a nozzle at the other end. These simple motors, a mere 50 mm in diameter and referred to as RH-00 (the 'RH' standing for Rohini), were then fired.

The tests often resulted in the motor exploding. At the time, proper facilities where the test could be monitored remotely were not available. Undaunted, people sheltered behind the plentiful coconut trees during tests. Since the metal nozzle took time to fabricate, it was often tied with a long rope so that it could be easily retrieved should there be an explosion.

Early indigenous sounding rockets

The first indigenous sounding rocket was a RH-75. The '75' referred to the rocket's diameter in millimetres. The 75 mm diameter was chosen because seamless aluminium alloy tubes of this size were commercially available. The

An RH-75 on its launcher

Cordite Factory at Aravancadu in Tamil Nadu provided the solid propellant needed in the form of cordite blocks. Cordite is a mixture of nitroglycerine and nitrocellulose. Nitroglycerine is the principal ingredient in dynamite which Alfred Nobel invented in 1864. Cordite came in the form of cylinders with a hollow core. Packed in boxes, the cordite cylinders were transported to Thumba in a jeep over bumpy, potholed roads. Considering that cordite is explosive,. it is surprising how few precautions were taken. 'We were young those days and didn't know any better,' one person commented.

RH - 75

Length : 1 metre

Diameter : 75 mm

Launch Weight : 7 kg

Propellant Weight : 2.5 kg

Propellant burn time : 2 seconds

Height reached : 7 km

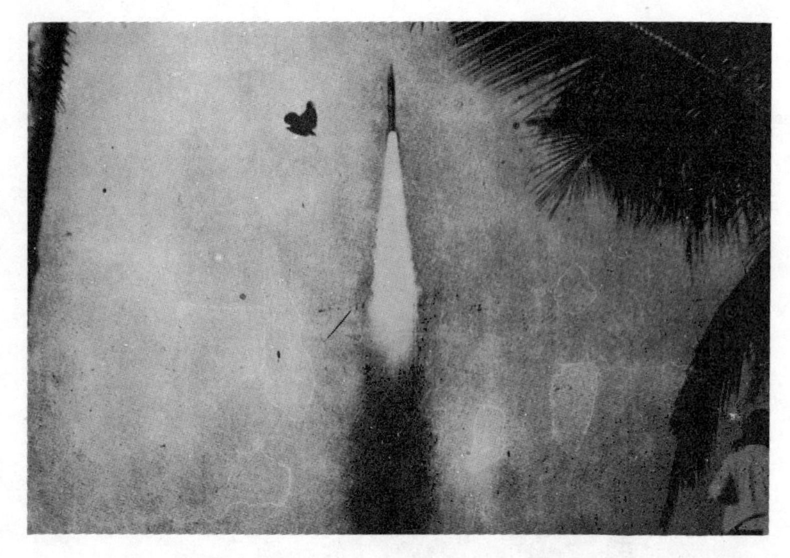

RH-75 soon after lift-off

Back in Thumba, these cordite blocks were slid into the aluminium casing. The casing was closed at one end and was fitted with a simple igniter of black powder. Above this was the nose cone. The RH-75 was too small to carry any payload. At the lower end, a nozzle was fitted. There were four fins on the outside. The first indigenous sounding rocket was probably launched on 20 November 1967. Its cordite propellant burnt for just a couple of seconds.

This early effort is said to have been carried out by a small team headed by A.E. Muthunayagam. Muthunayagam would later assume responsibility for the key area of liquid propulsion and create the Liquid Propulsion Systems Centre which develops liquid engines for satellites as well as launch vehicles.

Subsequently, the RH-100 and RH-125 sounding rockets were built and launched with cordite propellant. The RH-

125 was the biggest of the cordite-based Rohini sounding rockets. Its diameter of 125 mm was dictated by the size of the largest aluminium alloy tube available at the time. It carried the biggest cordite blocks which Aravancadu could make.

Even with these bought materials, innovation was necessary. Cordite blocks had to be bonded together to get the necessary length and their central bore enlarged by careful drilling. Later on, the Aravancadu factory was persuaded to produce cordite blocks with a star-shaped central bore specially for sounding rockets, instead of the usual circular bore. Thumba made and supplied special dies which Aravancadu needed to make these cordite blocks. Compared to a circular bore, solid propellant with a star-shaped bore has greater surface area exposed to burning. Larger quantities of hot gases were generated and resulted in greater thrust. When it turned out that the metal of the motor casing was too thin to take a thread, the nozzle was pushed to fit tightly into the casing and then held in place with screws.

The use of cordite propellant was, however, only the first step. The race was on to be the first to fire a Rohini rocket with a propellant developed in-house. Two ambitious young men, who would remain lifelong rivals, were both equally determined that their group should achieve that distinction. One was Muthunayagam. The other was Vasant Gowariker, a chemical engineer who was working on tactical missiles at the Summerfield Research Station in Britain when Sarabhai recruited him. Gowariker would go on to become director of the Vikram Sarabhai Space Centre, India's leading institution for launch vehicle development. Both are men who could have become chairmen of the

Indian Space Research Organization. Gowariker left the space programme after U.R. Rao became chairman and went on to head the Union government's Department of Science and Technology. Muthunayagam left when K. Kasturirangan got the top job in the space programme. He became head of the Union government's Department of Ocean Development, a position he currently holds.

It may look odd that two independent groups should both be working on indigenous solid propellants. But, as pointed out in the previous chapter, Sarabhai didn't believe in hard and fast rules regarding territories. In fact, he encouraged competition, probably taking the view that such competition would spur creativity as well as hard work, and lead to success. So if someone wanted to develop something, that person was usually allowed to go ahead even if others were working on exactly the same thing.

Muthunayagam's group decided to base their solid propellant around a natural rubber resin which was commercially available. Gowariker's group chose a solid propellant formulation based on polyester, which too was commercially available and used for production of plastics. This polyester-based propellant formulation was named 'Mrinal', after Sarabhai's wife, Mrinalini. Not surprisingly, accusations of currying favour with the boss ensued. The RH-75 with the Mrinal propellant was named the Dynamic Test Vehicle (DTV).

Versions differ on which of the two RH-75 projects produced the first indigenous sounding rocket with propellant developed at Thumba. Indeed, even today, people who then belonged to one group question whether the other RH-75 flew at all! Muthunayagam's bio-data states that the RH-75(S), with the natural-rubber propellant, was flight-tested in 1968. On the other hand, Gowariker's team claims that their DTV was successfully flown on 21 February 1969 and was 'the first Indian indigenous test rocket'. The anniversary of this launch has been observed as PED Day (PED being the Propellant Engineering Division which Gowariker headed) and there are moves to observe it as National Polymers Day.

As head of the Multi-Stage Rocket Project started in 1968, Muthunayagam was responsible for ISRO's early indigenous two-stage sounding rockets. The first of these, consisting of a modified RH-75 on top of a RH-125, was successfully flight-tested on 30 August 1968.

Making the Centaure rocket in India

As part of the contract with Sud Aviation to license-produce the Centaure two-stage sounding rocket in India, a small group was sent to France for training. On their return, this group, headed by M.R. Kurup, would establish the facilities for production of Centaures and be responsible for making these rockets in the country.

Kurup and others point out that the Centaure production facilities were not provided by the French on a turnkey basis. According to Kurup, the Indians were put through some general courses on solid propulsion. They were also given practical training in the fabrication of hardware for the Centaure and in casting the polyvinyl chloride (PVC) propellant these rockets used. In addition, there were visits to French solid propellant facilities.

The Rocket Propellant Plant (RPP), where the PVC propellants for Centaure would be cast, was set up right next to the Thumba launching station. According to Kurup, the French provided the list of equipment and their manufacturers as well as the designs of toolings required for making the hardware. The Indians had to design the buildings, keeping in mind safety requirements, based largely on what they had observed at the French facilities. They also had to install and commission the equipment. It turned out to be an invaluable experience.

The rocket motor casing and other metal parts for the indigenous Centaure rockets were initially fabricated at the Central workshop of the Bhabha Atomic Research Centre in Trombay. Two of the first three Centaures made at this workshop were filled with imported propellant. One of those rockets was launched from Thumba on 26 February 1969 with an Indian payload to measure the rocket's performance. The rocket reached a height of 145 km and its 31 kg payload also included a scientific experiment for measurements in the upper atmosphere.

February 1969 also saw the commissioning of RPP. The first ground test of the RPP propellant was successfully carried out on a 40 kg propellant block on 2 March 1969. Subsequently, a full-size Centaure booster was also made and ground-tested. The first indigenous Centaure with propellant made at RPP flew on 7 December 1969. Apart from a scientific payload and instrumentation to measure the vehicle's performance, it also carried the first Indian transponder which emitted radio signals so that the rocket could be tracked easily by radar. This rocket reached a height of 125 km. The following year, the Rocket Fabrication Facility was ready and all further work on sounding rockets could now be carried out at Thumba itself.

The successful indigenization of Centaure technology was a major landmark in establishing sounding rocket capability. The natural-rubber propellant and the Mrinal propellant were useful in proving that the technology for solid propellants could be successfully developed in India. But it is doubtful whether either of these propellants flew in more than one RH-75. By contrast, the PVC technology became the mainstay of the Rohini sounding rocket programme till better solid propellant formulations became available.

The Mrinal, like the Centaure's PVC propellant, is categorized as a composite propellant. Cordite, on the other hand, is what is called a double-base propellant in which the same chemical acts as fuel and oxidizer, providing oxygen needed for the fuel to burn. Double-base propellants, which continue to have their uses, have poor energy efficiency. In practical terms, this means that the thrust they provide per kg of propellant burnt per second, termed specific impulse, is not very high. Double-base propellants are therefore unsuitable for advanced rocket applications.

Composite propellants, where the fuel and oxidizer are separate, were a later development. As the name suggests, a number of ingredients go into the making of these propellants. Polymers, such as those used to make plastics, act as binders, holding the other constituents in place and setting them into solid blocks. These polymeric binders also act as fuel. Ammonium perchlorate is usually the oxidizer. A metal powder, usually aluminium powder, provides further fuel and, by raising the combustion temperature, also increases the thrust generated.

The difference between Mrinal and Centaure's PVC technology was that one was little more than a laboratory demonstration and the other was a mature production capability. Just a few kilogrammes of Mrinal were needed for the simple DTV. By contrast, each Centaure needed well over 300 kg of propellant. While just one small sounding rocket with the Mrinal was tested, over fifty of the Centaure-IIBs flew. Many indigenous Rohini sounding rockets used the PVC propellant till more advanced formulations became available.

The Centaure technology provided vital understanding of equipment, facilities, procedures and safety precautions needed to make large solid motors. Thirty years after it was set up, the Rocket Propellant Plant, expanded and improved, is still used to produce solid motors.

The Centaure technology also brought with it the use of 15 CD V6 steel, more advanced fabrication techniques and experience with two-stage rockets.

The tremendous strides the space programme would make in solid propulsion over the next two decades were a result of its not being content with simply importing technology. Sarabhai's judgement in using the Centaure deal to upgrade the technological level in the country and simultaneously to encourage indigenous efforts paid off. The merger of experience gained from these different import and indigenous streams provided the impetus for self-sustaining technological growth in solid propulsion.

Development of other sounding rockets

Establishment of the Rocket Fabrication Facility (RFF) for making hardware and of the Rocket Propellant Plant for casting solid motors led to a whole family of sounding rockets being designed and built in rapid succession. The RH-100 and RH-125, which had initially been launched with cordite propellant, were subsequently flown with PVC propellant. The new RH-125 weighed 32 kg and could take a 7 kg payload to a height of 10 km. A two-stage sounding rocket made by stacking two RH-125s one on top of the other was tested in 1970.

The Menaka-I, which flew in 1968, and the Menaka-II, which flew two years later, too were two-stage sounding rockets developed for meteorological applications. Altogether some 134 Menaka rockets are said to have been launched.

Menaka-II

The development of the Rohini sounding rockets continued well into the Seventies and beyond. Rohini rockets were flown with newer and better propellant formulations. These included solid propellants based on polypropylene glycol and the indigenously developed HEF-20 used in India's early launch vehicles. Likewise, a solid propellant formulation based on another indigenously developed substance, ISRO polyol, was used in the RH-300 sounding rockets. The single-stage RH-300 rocket was first launched in January 1983 and lengthened a few years later to accommodate 100 kg more of propellant. The resulting RH-300 Mk-II became a replacement for the two-stage Centaure sounding rocket.

By adding aluminium powder to the formulation, the energy of the PVC-based propellant was increased. The aluminized PVC propellant was used for developing the two-stage RH-560. This rocket, modelled on the French Dragon sounding rocket, is the largest of the Rohini series. When it was first launched in April 1974, the RH-560 carried an 86 kg payload to over 280 km.

The RH-200 sounding rocket was developed for meteorological observations. The first stage of this two-stage rocket was made of a strip of high-strength steel welded spirally to form a tube. The result was a lighter motor case which improved the rocket's performance. The rocket could carry 10 kg of payload to a height of 65 km. Some 182 RH-200s were launched during the 1979 Monsoon Experiment (Monex). For Monex, apart from Thumba and Sriharikota, sounding rockets were also fired from a new rocket launching station established at Balasore in Orissa. Balasore became operational with the launching of a RH-200 in January 1979.

Apart from the helical welding technique used for the RH-200 booster, there were other attempts to reduce the weight of the motor casing. A 100 mm diameter rocket with fibreglass casing flew in 1969. Strips of shaving blade quality steel were wound and bonded together to produce motor cases from 96 mm to 560 mm in diameter.

A decade later, paper-wound rockets, just 125 mm wide and capable of taking 10 kg payloads to a height of 5-6 km, were tested. These rockets could be destroyed in mid-air by detonating explosive charges onboard without the danger of creating metallic debris. They could be launched from a populated area and be used to disperse chemicals for hail suppression or as rocket flares to provide warning (say, of an approaching cyclone). There is nothing to indicate, however, that these paper-wound rockets went beyond a few test flights.

Even before the Rohini sounding rockets came on-stream, efforts to build instruments and payloads within

India had begun. Initially, they were sent on foreign sounding rockets, such as the Nike-Apache supplied by the United States. These payloads made it possible to track winds at various heights or measure different atmospheric parameters as the rocket ascended. Ways of ejecting or opening the payload fairing at specified heights were developed, as also the deployment of long booms with experimental probes. Electronics to radio down data and to emit radio signals for tracking the rocket accurately were tested and flown. Radio signals could be also be used to send commands to the rocket while it was in flight.

There were, of course, problems too, some of them hilarious. A few rockets exploded on the launch pad. In one, the nose cone and payload fell off just before launch and the rest of the rocket sped off into the sky. Another went straight into the ocean and, amazingly, came right out again, leading to jokes about it being the country's first 'sea-launched rocket'. One rocket, instead of taking the usual trajectory over the sea, veered so sharply off course that it headed inland and its payload was reportedly retrieved near the Trivandrum Engineering College.

The effort begun in the mid-Sixties to build sounding rockets within the country had reached a reasonable degree of maturity in the early Seventies. It created a base from which the more difficult challenge of building a launch vehicle could be attempted. These sounding rockets also became test beds to try out key technologies required for launchers. An RH-560, for instance, became the platform to test the guidance and control system for the SLV-3, India's first launch vehicle.

But the growing sophistication of satellite-borne sensors rang the death knell for sounding rockets. These satellite sensors could collect information globally and repetitively at a much lower cost than sounding rockets. During Monex, sounding rockets were launched from Thumba every week. They are rarely launched these days. Thumba can still make the RH-200, RH-300 Mark-II and the RH-560 Mark-II sounding rockets, but does so only after an order is received for them. An RH-300 Mk-II was supplied to the Norwegian Space Centre in 1997.

CHAPTER

3

SLV-3: India's First Launch Vehicle

SARABHAI WAS QUITE clear in his mind about India's need for an independent launch capability. There was, as he said at the dedication of TERLS to the United Nations, 'no ambiguity of purpose'. Applications in remote sensing, communications and meteorology made possible by satellites were directly relevant to India's fundamental problems. India had to be able to build applications satellites and also have the capability to launch them.

In 1968, when he addressed the United Nations Conference on the Exploration and Peaceful Uses of Outer Space, Sarabhai spoke about the importance of indigenous capability. While acknowledging that the advanced nations had done much to extend the benefits of space research to other countries, he pointed out that there were complex political implications when a country had to depend on another for launching its satellites. As long as there was no effective interdependence between the two, many nations would prefer to have some launch capability under their own control.

Sarabhai understood the problems which countries wanting to establish such launch capability would face. 'The military overtones of a launcher development programme of course complicate the free transmittal of technology involved in these applications', he observed in his speech. He went on to remark that knowledge could not be contained within artificial boundaries. Countries which possessed the technology had to learn to control through sharing knowledge, not by withholding it.

So, from his various published statements and verbal ones substantiated by those close to him, it is clear that, in Sarabhai's view, India could not afford to be dependent on other nations for either its satellites or the capability to launch them. Even today, this remains the rationale of the Indian space programme. It is equally evident that Sarabhai was well aware that other countries would be wary of India embarking on a launch vehicle development because of its potential missile applications, and technology would not be readily available. He nevertheless decided to go ahead.

Finding a new launch site

In fact, some months before he made that speech at Vienna, Sarabhai had initiated the first concrete steps which would culminate in India's satellite launch vehicle. After the ceremony in February 1968 to mark the dedication of TERLS to the United Nations, Sarabhai met with the space scientists. Pramod Kale remembers him saying that it was time for a full-fledged feasibility study on developing a satellite launch vehicle. The task of producing such a report, according to Kale, was handed to Y.J. Rao and himself.

The search began for a suitable launch site on the east coast. A rocket launched eastwards would be able to take advantage of the earth's rotation to give it an additional push. In addition, the spent stages of the launcher would fall into the sea where there is much less chance of their causing loss of life or property. It is for this reason that many launch sites in the world, including Cape Canaveral in the United States, the Centre Spatial Guyanais in French Guiana used by the Europeans, and Japan's Kagoshima launch complex are located on the eastern seaboard.

According to Kale, in early 1968, an offer came from the Andhra Pradesh government, suggesting the island of Sriharikota. After meeting the state government officials at Hyderabad, E.V. Chitnis and Kale went down to see the place for themselves. Sriharikota, which lay some 80 km north of the city of Madras (now called Chennai), turned out to be a sliver of land separating the brackish waters of the Pulicat Lake, the second largest lake in India, from the Bay of Bengal. The island has a coastline of 62 km, an area of 170 sq. km, and is no more than 8 km at its widest. It is largely scrub land, with plenty of eucalyptus and casuarina trees, teeming with snakes and mosquitoes. It gets blazingly hot in summer. The island was then sparsely inhabited, with only Yanadi tribals and a few scattered hamlets to be found there.

Kale spent the next three or four months surveying the entire stretch of coast from north of the island down to Chidambaram in the south. It would have been advantageous to have the launch site as close to the equator as possible. There was no point looking at sites south of Chidambaram, says Kale, as these would have the risk of the spent stages falling on Sri Lanka. Sriharikota is 13 degrees north of the equator and, at present, only the European launch complex in French Guiana is closer to the equator.

In August 1968, Kale submitted his report, recommending Sriharikota for the launch site. The *1968-69 Annual Report* of the Department of Atomic Energy states:

> Due to insufficient space at the Thumba range and due to limitations imposed by range safety conditions, a second rocket launching station is necessary to cope with the increasing rocket launching schedules at TERLS and SSTC. A station located on the east coast of India is necessary for facilitating a satellite launch. Sriharikota Island situated north of Pulicat in Andhra Pradesh has been found most suitable for this purpose.

R.D. John, the chief engineer for civil constructions, recollects what a singularly remote and isolated place Sriharikota was then. There was no proper road, only a mud track which would get washed away in the monsoon. One had to take a boat and cross the Buckingham Canal to reach a guest house run by the Forest Department of Andhra Pradesh. The journey from Sullurpeta, the nearest town, easily took an hour to an hour and a half. On the island itself, the best way to get from the guest house to the coast itself was to use the Forest Department's inspection trolley. Pushed by two labourers, it ran on narrow gauge rail tracks intended to transport forest timber.

When Sarabhai visited Sriharikota in May 1969, a temporary track was made up to Sriharikota by laying some bushes, casuarina branches, palmyra leaves and other materials on top to give the jeep tyres some grip. Six jeeps transported Sarabhai and the people who accompanied him to Sriharikota. John recalls how the friction between the jeep tyres and the sand heated the casuarina leaves to the point that they caught fire. This set fire to the jeep's tyres and the only way to extinguish the flames was to throw sand over them.

In February 1969, when the decision to set up a sounding rocket launch station was taken, the Andhra Pradesh government made the land available free. A temporary track was built over the Buckingham Canal and the construction material transported by bullock-cart and jeeps. The *Annual Report for 1969-70* says:

> Shar has an area of about 12,141 hectares having a 20 km coast line and approximately 10 km maximum width. It is a forest area without any permanent habitation or cultivation except eucalyptus and casuarina trees used mainly as firewood . . . Out of the total area of 12,141 hectares earmarked by Andhra Pradesh for this project, an area of approximately 3,845 hectares lying within a radius of 5 km, has already been handed over by the Andhra Pradesh Government to the Project authorities. The work on the lay-out of facilities has commenced. It is expected that in six months time a sounding rocket launching facility would be put up and used for flight testing of new rockets being indigenously developed at the Space Science and Technology Centre in Thumba. Later, facilities would be added as required for satellite launching during the next three years.

The Sriharikota Range became operational with the firing of a RH-125 sounding rocket on 9 October 1971.

The decision to build a launch vehicle

A month or so after Kale submitted his report on Sriharikota, the report on the feasibility of building a launch vehicle was also ready. According to Kale, the report took the view that a four-stage launcher, modelled on the Scout of the United States, was practical in the Indian context. The *1968-69 Annual Report* remarks that 'the TERLS and SSTC engineers have prepared a preliminary study report for launching a modest scientific satellite of about 20-40 kg weight in 400 km orbit'.

Y.J. Rao was asked to do a more detailed study of possible configurations for the launch vehicle. Rao and his team reportedly came up with four configurations — according to another version, there were six configurations — all of them involving four solid stages, differing only in diameter and length. These configurations were probably numbered SLV-1, SLV-2 and so on, with SLV standing for Satellite Launch Vehicle. So when Sarabhai picked the third configuration, the name SLV-3 stuck. This may have occurred in 1969. According to the Department of Space's *Performance Budget for 1975-1976*, 'in 1970, a decision was taken to undertake the indigenous development of a satellite launch capability for placing a scientific satellite of 40 kg in a 400 km near-circular orbit around the earth'. The SLV-3 was to be got ready in eight years at a cost estimated at Rs 15.6 crore.

In 1969, an important organizational change occurred. In 1962, the Department of Atomic Energy had created the Indian National Committee for Space Research (INCOSPAR). With Vikram Sarabhai as its chairman, this committee was responsible for space research and development. On 15 August 1969, INCOSPAR was reconstituted under the Indian National Science Academy, the national body affiliated to the Committee on Space Research (COSPAR) of the International Council of Scientific Unions. Vikram Sarabhai, who had by then taken

Bhabha's place as head of the atomic energy programme, created the Indian Space Research Organization (ISRO) as the apex body governing the space programme. He was its chairman till his death. To this day, the head of the space programme is always best known as chairman of ISRO. The practical relevance of having a body such as ISRO will be discussed in a later chapter.

The challenge of developing a launch vehicle

Sarabhai must have had extraordinary faith in his people to have seriously considered building launch vehicles in 1968. The demonstrated Indian capability even in sounding rockets at the time was quite modest. After all, a completely indigenous Centaure rocket, the first modern two-stage sounding rocket built in the country, flew only in December 1969.

By launch vehicle standards, the SLV-3 is a simple launcher. Yet an enormous technological gulf separates it from sounding rockets. A sounding rocket is smaller, much lighter and infinitely less complex. Its task is only to keep rising vertically till its propellants are exhausted, after which it falls back to earth. A launch vehicle, on the other hand, has to inject a satellite at the right height, angle and speed. Only then will the desired orbit be achieved.

As ISRO's own experience shows, sounding rockets can be made by a small group of engineers. But to build even the simplest launch vehicle is a very different story. A number of complex systems have to work together for a launch vehicle to succeed. As a result, any launch vehicle requires the activities of multi-disciplinary teams being coordinated and focused. Project management for the development of a launch vehicle, from conception to execution, is therefore a complicated exercise in itself.

No single stage can by itself put a satellite into orbit. So a launch vehicle consists of a number of stages. Each stage has a propulsion system, either solid or liquid. When its fuel is exhausted, that stage is jettisoned and the next stage is ignited. The SLV-3, for instance, has four stages. The stages are optimized as far as possible to reduce their inert weight and thereby improve their performance. The stages also have systems to control the launch vehicle's orientation and prevent it from tumbling out of control. In flight, each stage of a launch vehicle has to perform exactly as planned if the mission is to be successful.

Although a sounding rocket might have two stages, these stages do not have to be so highly optimized as in a launch vehicle and their performance in flight is less critical. A sounding rocket is kept stable in flight with fins. In addition, it is set spinning soon after leaving the launcher and this spin adds to its stability. The fins and spin stabilization suffice to keep the sounding rocket ascending vertically. There are no mechanisms to correct deviations from the trajectory.

A launch vehicle, on the other hand, has to be guided along a curving trajectory to the precise point where the satellite will be injected. Onboard sensors and guidance systems are needed to determine the satellite's position and suitably adjust its course. On command from the guidance system, the control systems have to change the vehicle's orientation when needed, either to correct a disturbance or to steer the vehicle. Events such as separation of spent stages and ignition of the next stage have to be carried out at precisely the right moment.

As it flies along the trajectory which will put the satellite into orbit, the launch vehicle is subject to enormous stresses and strains. It, therefore, has to have a structure strong enough to withstand these stresses. On the other hand, if the structural elements are made too strong, their weight will extact a price in terms of the payload that the launch vehicle can carry. Extensive wind tunnel tests and a large number of studies, including, nowadays, computer simulations, are carried out to understand the airflow patterns around the launch vehicle during its flight, to judge whether these could cause problems, and also to determine the kind of loads which the launch vehicle would be subjected to.

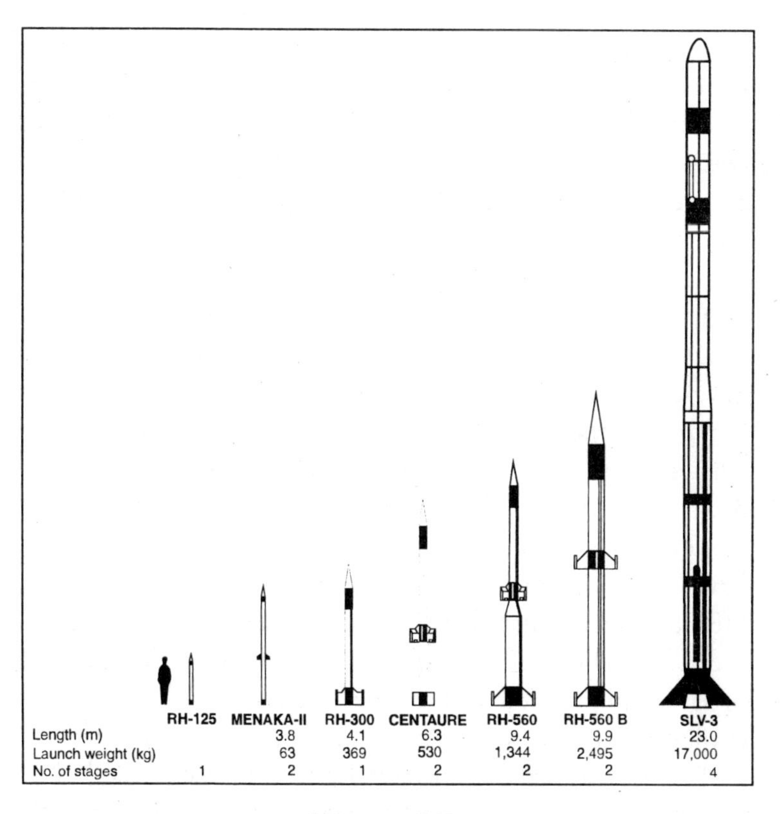

RH-125 to SLV-3

The Centaure-IIB, the largest of the ISRO-built sounding rockets before the SLV-3 programme began, had two stages, was 6.3 metres long and weighed just 530 kg. The SLV-3, quite a small and simple launch vehicle, had four stages, stood 23 metres high and weighed 17 tonnes at lift-off. The propellant alone weighed 13 tonnes. It was made up of 44 major sub-systems, had nearly 7,000 electrical components, a million soldered joints, 25 km of wiring and 40,000 fasteners. The transition from making sounding rockets to building launch vehicles is not an easy one.

The SLV-3 configuration

The detailed design and development of each of the four stages of the launch vehicle was handled as a separate sub-project, each by a different project leader. Gowariker headed Design Project Stage-1 (DPS-1), Muthunayagam was in charge of DPS-2, Kurup of DPS-3 and Kalam of DPS-4. S.C. Gupta was given charge of the design project for inertial systems and guidance. There is also said to have been another design project dealing with telemetry.

It is not clear why Sarabhai chose to run the SLV-3 development programme with separate project leaders. Maybe he thought that with all the personal rivalries in existence, picking out any one person from the space programme and making him project director for SLV-3 would only cause dissatisfaction and more feuding. He may have drawn inspiration from the European effort to build the Europa launcher, with the British, French and Germans each contributing one of its three stages. (The Europa programme turned out to be a failure and its project management may well have been responsible for this.)

But having separate project leaders for each SLV-3 stage also meant that much of the overall responsibility for the programme fell directly on Sarabhai, an onerous task for a person in charge of the country's atomic energy programme. Many believe that if Sarabhai had lived longer, he would have had to appoint a single project leader for SLV-3.

When configurations for a satellite launch vehicle were considered, it was quite clear that solid and not liquid engines would have to be the primary source of propulsion. The reason quite simply was that the Indians were more advanced in solids than in liquids. Although some work had begun in liquids, it was still fairly rudimentary. By 1969, the Rocket Propellant Plant (RPP) was operational and Indian-made Centaures were flying. RPP was expanded to cope with making solid motors for a launch vehicle. No comparable capability existed on the liquid side. It would be 1973 before a sounding rocket with an upper stage powered by a liquid engine could be tested.

An additional factor for using solid engines may have been Professor Hideo Itokawa of the Institute of Space and Aeronautical Sciences in Japan, whom Sarabhai had appointed as a consultant. The Japanese had orbited their first satellite, the 23 kg Osumi, in early 1970 using the Lambda-4S launcher which had four solid stages as well as two solid strap-on motors. Since the technologically more advanced Japanese had opted for the solid route, it would have reinforced opinion in India that this was the way to go.

More important, the Indians had in mind a launcher on which they could model their own launch vehicle — the Scout developed by the United States. An acronym for Solid Controlled Orbital Utility system, the Scout was an all-solid four-stage vehicle. Intended as a low-cost launcher for scientific satellites, its stages were derived from ones developed for American military programmes. The first

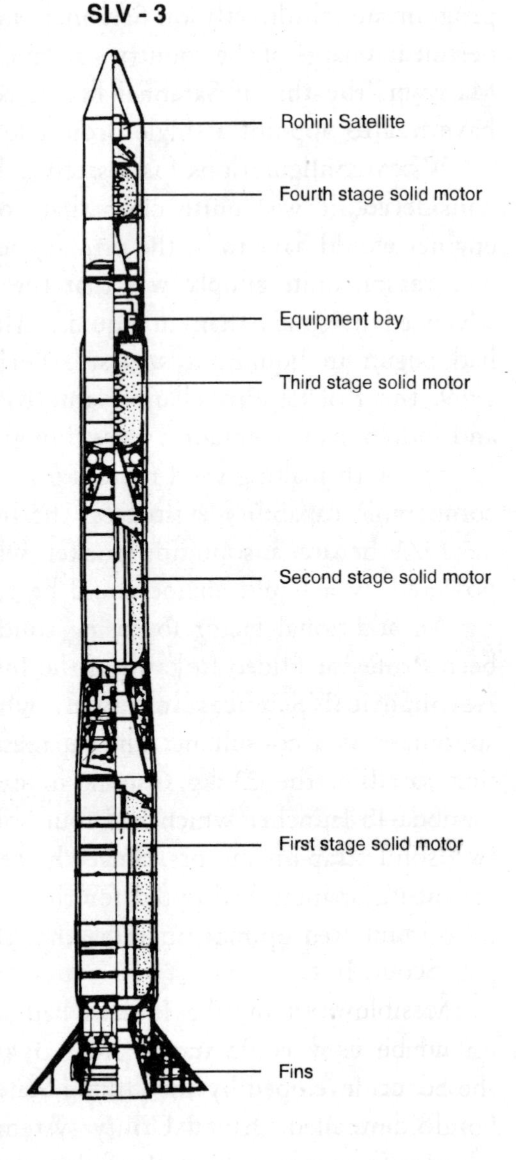

stage motor was derived from the Polaris missile, the second stage from the Sergeant missile, and its upper two stages adapted from the Vanguard rocket. The Scout first flew in July 1960 and, with over a hundred launches since then and a success rate of 96 per cent, it had earned a reputation for reliability.

The Scout was launched from NASA's Wallops Island facility which was also used for sounding rocket launches. It was at Wallops Island that the small group sent from India had, in early 1963, received training with a view to establish a sounding rocket launch station at Thumba. These men must have known about the Scout and perhaps also something of its design. So, when Sarabhai decided it was time for India to build a launch vehicle of its own, the Indian design was closely modelled on the Scout. This was probably done because the Indians did not then have sufficient experience for *ab initio* design of a launch vehicle. As with any aerospace system, whether a passenger aircraft or a launcher, designers have to be able to predict accurately how a particular configuration will behave in actual flight, compute the various forces and stresses which will act on it, and make sure that the vehicle is able to carry out its mission. Such capability is based on actual flight experience combined with extensive wind tunnel testing and mathematical models which can predict what will happen. That capability would be built up in ISRO only during the SLV-3's development.

Since the Scout had successfully flown, the Indian scientists could be confident that a four-stage wholly solid launch vehicle of this sort was definitely feasible. Knowledge of the dimensions and other characteristics of the Scout stages permitted them to fix the performance requirements for the SLV-3 stages with, of course, appropriate allowances for Indian inexperience.

While knowledge of the Scout simplified the task of deciding the SLV-3's configuration and design, it is important to realize that the SLV-3 is not a mere copy of the Scout. In fact, such a thing would have been completely impossible without licence manufacture. While the Scout could be used as a reference, the SLV-3 design had to be based on materials, propellants and systems the Indians could develop or get hold of. So the SLV-3 could not have been built by blindly copying Scout. ISRO engineers needed to understand the principles underlying the Scout design and then apply them to the Indian situation.

Moreover, design is only the beginning. Technology had to be developed to meet those design specifications. Every single technology element, from propellants to the electronics, had to be developed, tested and proven to be reliable enough to be used in a launch vehicle. Just how difficult the task turned out is shown by the fact that an entire decade – all of the Seventies – went in developing the SLV-3.

Launch vehicle development requires specialist teams in a number of different disciplines, sophisticated facilities for analysis and for ground testing as well as considerable fabrication and production capabilities in-house and in industry. The number of scientific and technical personnel involved quadrupled during the SLV-3 period. Equipment was made or imported. Existing facilities were augmented and new ones, including solid propellant casting and testing facilities as well as the launch complex at Sriharikota, built. The launch vehicle programme consumed 70 to 90 per cent of the total space budget during this period.

Since the Indian engineers didn't just copy the Scout, there are significant differences between the Scout and the SLV-3. Using solid motors from US military missiles, the Scout had better optimized solid stages. The lower inert weight in the Scout stages gave it significantly better performance. As it was their first launcher, ISRO was less ambitious.

The SLV-3 first stage was made up of three segments instead of the single monolithic motor used in the Scout. The reason for this will be examined in the next chapter. As in the early Scouts, the upper two stages of the SLV-3 had fibreglass casing. If metal had been used for the casing, the greater weight of the casing would have reduced the performance of the motor. In the fourth and final flight of the SLV-3, the fourth stage casing was changed to Kevlar, a high-strength fibre. This made it possible to reduce the weight of the motor casing still further. The Scout's fourth stage had moved to Kevlar casing earlier.

Although solids propellants provided the propulsion for all four SLV-3 stages, control systems were needed to maintain the vehicle in the correct orientation and to guide it along a predetermined trajectory. In the second and third stages, small engines using liquid propellants, called thrusters, were used for the purpose. There were differences between the Scout and the SLV-3 in the control systems used.

The SLV-3, like the Scout, had a simple guidance system. The development of inertial sensors and guidance is discussed in a later chapter. Access to better sensors and greater experience provided Scout with a better guidance capability.

Death of a visionary

On the morning of 30 December 1971, ISRO woke up to the news that Vikram Sarabhai was no more. It was a terrible shock. He had flown down to Trivandrum, as was his wont, and spent the previous day in hectic meetings which went on into the night. Despite pressures on his time, Sarabhai had always tried to maintain direct contact with his technical staff. This trip was no different. There were people who wanted to show what they had achieved, others with proposals and ideas. Invariably, with a major project like SLV-3 in hand, there were all sorts of issues which had to be sorted out. As a result, a great many people had spent time with him. Now they could not believe that the man whom they had met just the day before, vibrant and energetic had been struck down in his sleep. The long hours he had worked and the heavy responsibilities he shouldered had taken their toll. Sarabhai was just fifty-two.

Soon, on the other side of the globe, an Indian professor of aeronautics would be called out of a class he was taking at the California Institute of Technology (Caltech) by a telephone call from the Indian embassy in Washington DC. The Prime Minister, Indira Gandhi, wanted him to return to India and take charge of the space programme. The professor was Satish Dhawan.

Born on 25 September 1920 in Srinagar, Dhawan graduated from the the University of Punjab with quite a combination of degrees — a BA in Mathematics and Physics, an MA in English Literature and a BE in Mechanical Engineering. He then went on to take an MS in Aeronautical Engineering from the University of Minnesota and a doctorate from Caltech. Dhawan then joined the Department of Aeronautical Engineering at the Indian Institute of Science as a senior scientific officer in 1951. He was professor and head of that department in just four years and, seven years later, the director of the institute itself. At forty-two, he was the institute's youngest director and when he retired eighteen years later, he became its longest serving one.

After having been the director for nearly nine years, Dhawan persuaded the institute to let him take a year's sabbatical. He wanted to spend 1971-72 at his alma mater, Caltech. Not long before the call from the embassy, Dhawan had received a letter from Sarabhai, asking whether he could assist Muthunayagam, who was in the United States looking for help, in setting up a solid motor testing facility. Dhawan was able to arrange for both of them to visit the Jet Propulsion Laboratory's test facilities.

Most people, given a request from no less a person than the Prime Minister to head the space programme, would probably have been on the next available flight to India. Not Dhawan. He told whoever called from the embassy to please tell the Prime Minister that he was on sabbatical and was, moreover, teaching a course. He would be able to return only after he had finished that. Secondly, as he was working for the Indian Institute of Science and was responsible to its council, he would be unable to give any firm answer for anything until he had talked to them.

The Prime Minister and the government, to their credit, decided to await Dhawan's return. In the interim period, M.G.K. Menon, who was then director of the Tata Institute of Fundamental Research and secretary of the Department of Electronics, was asked to head the Indian Space Research Organization as well. Menon became chairman of ISRO in January 1972.

After he came back to India, Dhawan went and met R. Choksi, chairman of the Indian Institute of Science's council, and J.R.D. Tata, president of the institute's court. Both of them made it clear that the institute would not stand in Dhawan's way and it was up to him to decide what he wanted to do. But Dhawan himself did not want to take even leave of absence, let alone resign from the institute.

Dhawan went to Delhi and met the Prime Minister. He was willing to head the space programme, but on two conditions. The first was that he should be allowed to continue as director of the institute. To enable him to do so, the second condition was that the headquarters of the space programme would have to be in Bangalore. Indira Gandhi promptly agreed to both conditions.

On 1 June 1972, the government, through a resolution, established a Space Commission modelled on the Atomic Energy Commission and also a separate Department of Space. The Indian Space Research Organization was brought under these. In September 1972, Satish Dhawan became chairman of the newly established Space Commission, secretary for the Department of Space and chairman of ISRO. The three posts would henceforth go together.

In the initial months after Dhawan assumed charge, ISRO headquarters operated from the Indian Institute of Science. Later space was found for it in Cauvery Bhavan on Bangalore's Kempe Gowda Road. Even then, many meetings with Dhawan continued to be held in the spacious and quiet offices of the director of the Institute. Dhawan remained director till his retirement from the institute in 1981. He continued to head the space programme till the end of September 1984 and has been a member of the Space Commission since then.

Sarabhai had believed in direct interaction. He had created several, almost autonomous, technical groups in both Trivandrum and at Ahmedabad. All proposals were presented to him and he alone decided whether or not any should be pursued. This method, and the sense of competition it created, had been useful in creating basic technological competence.

When Dhawan took over, three major tasks stood before ISRO. Its agreement with NASA for the loan of their ATS-6 satellite committed the organization to getting the ground hardware, such as the direct reception TV sets, ready in time, along with several hundred hours of programme content. Then, there was the SLV-3. In addition, in May 1972, during M.G.K. Menon's stewardship, an agreement had been signed with the Soviet Union for launching India's first satellite. This scientific satellite, later named Aryabhata, had also to be built.

In place of the one-to-one relationship between the ISRO chairman and technical groups of the Sarabhai period, Dhawan wanted organizational structures which could operate independent of the individuals involved. Dhawan decided to unify the groups, both at Ahmedabad and at Trivandrum. The Space Applications Centre (SAC) was formed at Ahmedabad. In Trivandrum, the Space Science and Technology Centre (SSTC), the Rocket Propellant Plant (RPP), the Rocket Fabrication Facility (RFF), the Propellant Fuel Complex (PFC), the Sriharikota Range (Shar), and the Indian Scientific Satellite Project were unified into the Vikram Sarabhai Space Centre (VSSC). Both the SAC and VSSC would have directors in overall charge. Dhawan persuaded Yash Pal, then a professor at the Tata Institute of Fundamental Research, to become the director of SAC. For VSSC, he approached Brahm Prakash.

Brahm Prakash came to VSSC with the reputation of being one of the country's best metallurgists. Born in Lahore of pre-Partition India on 21 August 1912, he had studied Chemistry at the Government College there. During this time, he caught the eye of Shanti Swarup Bhatnagar, then head of the Chemistry Department at the University of Punjab in Lahore. Bhatnagar later established India's Council of Scientific and Industrial Research (CSIR) and its chain of research laboratories. The Bhatnagar awards given annually by the government for outstanding contributions in science are named after him. Brahm Prakash took his M.Sc. and then Ph.D. in Physical Chemistry from the University of Punjab.

The paths of Brahm Prakash and Dhawan were to repeatedly intertwine. Both had studied at the University

L-R: P.D. Bhavsar, Satish Dhawan, Mrs Dhawan, M.R. Kurup and Brahm Prakash

of Punjab. They travelled together on the same ship to the United States, the former to take a post-doctoral degree (Sc.D.) in metallurgy from MIT and the latter for doctoral work at Caltech. On returning to India, Brahm Prakash was chosen by Bhabha to the position of metallurgist in the atomic energy establishment. However, as the atomic energy programme was just beginning, Brahm Prakash was deputed to the Indian Institute of Science, where he served as professor and head of the Department of Metallurgy for six years. He and Dhawan were neighbours at the institute.

In 1957, Brahm Prakash returned to the Atomic Energy Department and became director of the Metallurgy Group. Under his leadership, the group flourished and developed techniques for the extraction and fabrication of a variety of metals. Their achievements included the method for producing nuclear-grade zirconium, indigenously making the fuel elements for the Canada-India Research Reactor (CIRUS), and establishing the facilities needed to extract plutonium and fabricate plutonium fuels. He played a key role as project director in setting up the Nuclear Fuel Complex at Hyderabad, which today produces all the nuclear fuel needed by Indian reactors. Brahm Prakash was about to retire from Atomic Energy when Dhawan asked him to head VSSC.

Brahm Prakash, fondly known as 'BP', brought more to his new task than outstanding scientific credentials. A chain-smoker and a man of few words, Brahm Prakash was also a wonderful human being. As one person remarked, people respected Dhawan deeply, but BP was venerated. A 'saint' was invariably how people described him. These qualities stood him in good stead. Another person might have faced resentment from people and groups used to dealing directly with the ISRO chairman and being responsible only to him. But none of that occurred with Brahm Prakash. Instead, once a matter was discussed and a decision taken with his approval, it was considered settled. The success of SLV-3 is largely due to this quiet, self-effacing man. He stood above the petty squabbles and rivalries, and was able to make everyone involved in the launch vehicle development work together as a team.

Apart from unifying the groups at Ahmedabad and Trivandrum, the other important change Dhawan made was to unify the project management of SLV-3. Instead of having several design projects for the launch vehicle, with separate heads for each of them, Dhawan decided to have a single project director for the entire SLV-3. The need for one was obvious, says Dhawan. 'Everywhere in the world it is so. I didn't invent it. It seems to me that eventually whoever was going to run the programme would have to do it.'

The man chosen for the job was none other than Abdul Kalam. Kalam had been one of the earliest recruits to the space programme and a part of the initial group trained by NASA so that the Thumba launch facility could be established. On his return to Trivandrum, developing payloads for sounding rockets became one of Kalam's tasks. Kalam later initiated work into composites and was responsible for establishing the Reinforced Plastic Centre (REPLACE). When the SLV programme began, he was given charge of the design project (referred to earlier) of the fourth stage. Both the third and fourth stages of the SLV-3 would use fibre-reinforced plastics for their solid motor casing.

Dhawan felt that Kalam's abilities were not being fully utilized at REPLACE. 'In my opinion, he was being wasted there,' says Dhawan. Fibre-reinforced technology was based on a very difficult science and Kalam was not contributing to that. Kalam, he points out, had several interests, not just one, and was very good at making things. Another factor in Kalam's favour may have been his ability to get along with people. Dhawan sought Brahm Prakash's views. Brahm Prakash approved the choice and Kalam was appointed the project director for the SLV-3.

Kalam's was no easy task. In the pre-VSSC days, he was not a part of the inner circle. There were all kinds of subtle distinctions governing the pecking order in those days. There were those who had doctorates and those who did not have them, those who were a part of the core technology development groups at SSTC and those who were outside them. Kalam did not have a doctorate and was attached to TERLS, not SSTC. Neither was he a member of the Technical Coordination and Finance Committee which was nominally responsible for the running of SSTC. He was neither the seniormost nor the most powerful of scientists in Trivandrum at that time.

The support of Dhawan and even more of Brahm Prakash was crucial. In *Wings of Fire*, Kalam writes, 'he [Brahm Prakash] had always been my sheet anchor in the turbulent waters of VSSC'. The support did not come in the form of backing Kalam against others, for that would only have wrecked the teamwork essential to make the launch vehicle. Instead, committees were formed at various levels. A number of reviews too were conducted regularly, so that there was a constant check on work being done and bottlenecks could be removed. These committees and periodic reviews served the dual function of making it possible for the SLV-3 project team to coordinate the work being carried out by various groups, as well as giving those groups an important say in the decision making. This management system gave the project leader and his team considerable authority. But, at the same time, they did not have the carte blanche to do as they pleased since important decisions could only be taken after a wide-ranging consultative process during which every point of view raised had been considered. The systems also ensured tight adherence to schedules and budgets. This project management capability proved crucial for ISRO's later programmes.

SLV-3 is launched

Considering that SLV-3 was ISRO's very first launch vehicle, its development seems to have gone remarkably smoothly. But the task took a whole decade to accomplish, reflecting the magnitude of the challenge. (Details of the developments carried out in solid motors, liquid engines and inertial guidance are covered in later chapters.)

The considerable involvement of research institutions and industry in the development and making of the SLV-3 is noteworthy. Sarabhai had been emphatic about maximum utilization of the capabilities and facilities of other organizations and industry. Dhawan gave still greater impetus to ISRO going outside for its needs. The wind tunnels at the National Aerospace Laboratories (then the National Aeronautical Laboratory) and the Indian Institute of Science were extensively used. Both institutions were involved in a number of theoretical and experimental studies. Some forty-six major industries and institutions outside ISRO contributed to the building of the SLV-3. Larsen & Toubro and Walchandnagar had made motor cases. Hindustan Aeronautics Limited (HAL) carried out some of the fabrication needed. A number of small industries too were used.

Although the intention had been to have the first launch in 1978, a major mid-term review in December 1976 established that launch would be possible only in 1979. The first two launches were designated as 'experimental' launches and there would be two more classified as 'developmental' launches.

The first launch was called the SLV-3(E)-01, with the 'E' standing for 'experimental'. Its task was to put a 40 kg Rohini satellite into an elliptical orbit with the satellite coming no closer than 300 km to earth. In this orbit, the Rohini satellite would have a life of at least a hundred days. Like the Scout, the SLV-3 stages were integrated horizontally. The fully integrated launch vehicle, with the satellite enclosed within the heat shield, was taken to the launch pad and raised to the vertical position.

The first SLV-3 lifted off from Sriharikota at 7.58 a.m. on 10 August 1979. The first stage performed perfectly and separated without a hitch. But during the operation of the second stage, the launch vehicle began to deviate from the planned trajectory. Kalam poignantly conveys the disappointment they felt in his autobiography, *Wings of Fire:*

> Suddenly, the spell was broken. The second stage went out of control. The flight was terminated after 317 seconds and the vehicle's remains, including my favourite fourth stage with the payload splashed into the sea, 560 km off Sriharikota.

The flaw lay with the control system of the second stage. This system used a dozen thrusters, fuelled by red fuming nitric acid (RFNA) and hydrazine, to adjust the launch vehicle's orientation during the operation of the second stage. Electrically operated solenoid valves controlled the flow of propellants to the thrusters. A solenoid valve, controlling the flow of RFNA to one of the thrusters, failed to close and the nitric acid had leaked away. When the time came to activate the thrusters, there was no nitric acid left. With the thrusters inoperational, the vehicle went out of control and broke up.

Initially, it was believed that some contamination had resulted in the solenoid valve getting stuck. But analysis of data gathered during the countdown, parameters radioed down from the launch vehicle during its flight combined with extensive tests on the ground showed that contamination had not been the problem. It turned out that the electromechanical relays, which sent electrical power to the solenoid valves, had a design limitation. These relays could sometimes get fused when electrical power was drawn from the launch vehicle's onboard batteries. When this happened, the solenoid valve to which the relay was connected remained open and the propellant leaked out.

It later turned out that signs of the RFNA leak had been noticed when the thrusters of the second stage and subsequently those in the third stage were tested with onboard power during the countdown. While everybody else was inside the blockhouse, a couple of observers, equipped with binoculars, had been posted outside. They saw some dark fumes when the nitric acid leaked, but were puzzled as they soon stopped. Misled by what appeared to be only a transient phenomenon, it was decided to go ahead with the launch.

The *1979-80 Annual Report* of the Department of Space describes the SLV-3 launch as being 'partially unsuccessful'. It is an extraordinary phrase, reportedly put in by Dhawan himself, which has never again been used by ISRO. But it conveys with uncompromising honesty the fact that although most systems had functioned as planned, the SLV-3 launch had been a failure. Once the cause of the failure was established by the failure analysis committee, it was a relatively simple matter to modify the relay and carry out ground tests to make sure that the problem did not recur.

Little less than a year after the first failure, another SLV-3, designated SLV-3(E)-02, stood on the launch pad. On 18 July 1980, the day of the launch, tension was high at VSSC whose 4,000-odd employees felt a personal stake in the launch vehicle. A telephone line relayed the countdown at Shar and was broadcast over speakers in the VSSC auditorium, which was overflowing by six o'clock in the morning.

Countdown, the VSSC house journal, later recorded:

> The countdown which commenced 36 hours earlier was agonisingly suspenseful. There were many moments of anxiety which were fortunately brief and transitory. Except for a minor hold, the elaborate checks went perfectly smoothly.

The warm-up pulses for the second and third stage control systems went through flawlessly. Then at 8.03 a.m., the rocket was on its way. This time there were no problems of any sort. Twelve minutes and six seconds after launch, the SLV-3 put the 35 kg Rohini satellite into a 300 km by 900 km elliptical orbit. The satellite had been launched eastwards. After the satellite circled the earth and rose over India's western horizon, the first radio signals from the satellite were picked up at Trivandrum 1 hour and 45 minutes after launch.

VSSC and Shar exploded with joy, and the country celebrated with them. Kalam and Gowariker (who had taken over as director of VSSC when Brahm Prakash retired in November 1979) were mobbed at the airport by VSSC staff as well as the public when they returned triumphantly to Trivandrum.

But success brought problems for Kalam. In *Wings of Fire*, Kalam says that 'the unpalatable truth I had to face was that by becoming the focus of media attention, I had become the cause of bitterness among some of my senior colleagues, all of whom had contributed equally to the success of the SLV-3'. Matters were not improved when the following year's Repúblic Day honours brought a Padma Vibhushan for Dhawan and a Padma Bhushan for Kalam. Kalam seemed to be getting all the credit while others who had been instrumental in developing the necessary technology appeared to have been forgotten. Gowariker, for instance, would have to wait till 1984 for just a Padma Shri. Others didn't get even that.

Kalam later moved to the ISRO headquarters staff at Bangalore. In June 1982, he left ISRO altogether and went to head the Defence Research and Development Laboratory at Hyderabad, the principal organization for missile development. The man who replaced Kalam as SLV-3's project director was Ved Prakash Sandlas. After the SLV-3 project ended, Sandlas too moved to the Defence Research and Development Organization (DRDO) in September 1986 and ten years later became its chief controller for Research & Development.

The first developmental launch of SLV-3 took place on 31 May 1981. This time another problem surfaced. The first stage had finished firing but before it could be separated, the launch vehicle began spinning unexpectedly some 63 seconds into the flight. Despite this, the first stage separated without difficulty. Although the subsequent stages performed well, as *Countdown* reported, 'the unexpected spin continued to demand [its] pound of flesh'. The velocity when the satellite was injected into orbit was one per cent lower than planned and the injection angle about 0.5 degree off. This put the satellite into an orbit 183 km by 426 km which brought it so close to the earth that it re-entered and burnt up in the atmosphere nine days later.

The fourth and final launch of the SLV-3 came two years later, on 17 April 1983. This time a number of improvements had been made to the launch vehicle, the most significant of which was the Kevlar motor case for the fourth stage. The flight was uneventful and the Rohini satellite was placed in a 388 km by 852 km orbit. With it, the SLV-3 programme ended.

The significance of SLV-3

Considering that the SLV-3 was ISRO's first attempt at building a launch vehicle, it was quite an achievement. The Scout could draw upon the experience which the United States had accumulated in rocketry through years of civil and military programmes. Nevertheless, the first orbital launch of the Scout was a failure. Japan, which preceded India in launch vehicle development, suffered four consecutive failures in almost as many years with their Lambda-4S launch vehicle before putting their first satellite into orbit in February 1970.

The Scout's four stages were based on solid motors developed for the US missile and military launcher programmes. In case of the SLV-3, as we shall see in the next chapter, it was the first time that ISRO was attempting such large solid motors, and that too using modern high-energy propellant combinations. Despite this, the SLV-3 was able to carry a payload which was only 30 per cent lighter than that of the first Scout version.

The SLV-3 did more than make India one of the few countries capable of launching a satellite. Its real significance for the Indian rocket programme was that it gave ISRO the capability to develop the more powerful launch vehicles needed to put operational satellites into orbit. This, after all, was the whole purpose of the launch vehicle programme.

The goal for the Eighties would be to develop the Polar Satellite Launch Vehicle (PSLV) for launching the Indian Remote Sensing (IRS) satellites. As a stepping stone to test some of the key technologies involved, a simpler launcher, the Augmented Satellite Launch Vehicle (ASLV) would be attempted first.

CHAPTER

4

Developing Competence in Solid Propulsion

THE SOUNDING ROCKET programme created a base in solid propulsion. The Centaure and the other sounding rockets provided much needed experience in making solid motors with composite propellants. The Mrinal and the natural-rubber based propellant showed, if nothing else, that composite propellant formulations could be developed in India.

But much more needed to be done to meet the requirements of the SLV-3. New, higher energy propellant formulations had to be developed. Realizing that the chemicals required for making the propellants were a vulnerable choke-point, the ability to develop these chemicals as well as to produce them on a large scale was required. Apart from better propellant formulations, the SLV-3's solid motors had to have less inert weight so that their performance improved.

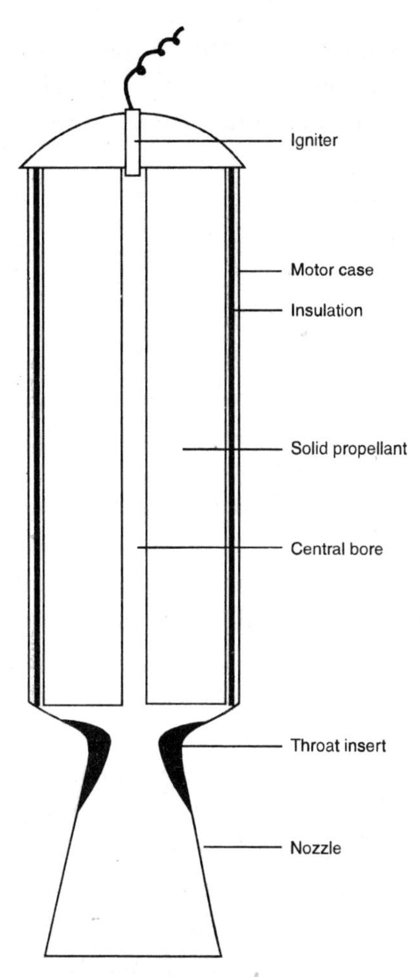

Typical Composite Solid Motor

The solid motors for the SLV-3 were bigger than anything ISRO had attempted before and facilities were needed both to make these motors and to test them. In addition, the problems of making flightworthy solid motors increase with their size and these issues too needed to be tackled. By the end of the SLV-3 programme, ISRO had first-rate capability in all aspects of solid propulsion.

At the heart of a composite propellant is the resin which acts as a binder. Chemically, it is a polymer, very similar to many plastic items of daily use. The resin, a viscous fluid, is mixed with other materials into a thick slurry. As the term binder suggests, the resin then sets into a rubbery mass which holds the solid ingredients in place, a process known as 'curing'. The resin also acts as fuel. One solid ingredient added is ammonium perchlorate, the oxidizer that provides the oxygen for the fuel to burn. Another solid ingredient added is the fine powder of a light metal such as aluminium. When the metal powder burns, it raises the temperature of combustion and thereby the energy imparted to the rocket. The cured block of solid propellant is termed the 'propellant grain'.

Solid propellant formulations vary to meet specific needs. Ammonium perchlorate typically accounts for 70 per cent of the weight of a composite propellant, aluminium powder for 15 per cent and the resin for only about 10 per cent. A variety of other chemicals added for different purposes account for the rest of the weight. Even though it makes up only a small proportion of the propellant, the choice of the binder affects the energy which a solid propellant formulation provides. The binder also determines the kind of defects the propellant grain is prone to in the face of various stresses, its reliability, the ease with which the propellant grain can be made as well as the storability of the grain and its cost.

The PVC-based propellant used in the Centaure series was outdated technology even at the time when it was acquired. With the SLV-3, ISRO moved rapidly into using the more advanced binders which had become available. In addition, for the solid motors needed for the SLV-3, it switched to case-bonded propellants. In the Centaure, the propellant slurry was poured into plastic tubes and, after it set, the tube, along with the propellant block it contained, was inserted into the motor casing. In case-bonded propellants, the slurry is poured right into the motor case itself. Before doing so, the motor case has to be lined with special materials for insulation and to improve the adherence of the propellant. The great advantage of case-bonded propellants is that inert weight is reduced and more propellant accommodated. Case-bonded propellants, therefore, give better performance.

Although solid motors appear simple, devoid of moving parts and the complex plumbing of liquid engines, they are also extremely unforgiving. A small defect can destroy the motor and the launch vehicle with it. These propellants are susceptible to stresses imposed on them during production and subsequent handling and operation. To make matters worse, solid propellants accumulate stresses. So defects such as cracks and voids can develop. The propellant can also get debonded from the casing. If such defects are large, the burning area increases suddenly when the motor is fired. The unexpected rise in pressure can then blow the motor apart.

Such defects can be avoided only by careful monitoring and by controlling every step of the process of making a solid motor. The difficulties of making a good quality propellant grain increase with the size of the motor. The solid motors required for the SLV-3 were much bigger than anything ISRO had attempted before. The launch vehicle's first stage would carry over 8 tonnes of propellant, the second stage 3 tonnes, the third stage 1 tonne and the fourth stage 262 kg of propellant. The solid propellants alone would account for three-quarters of the launch vehicle's total weight of 17 tonnes.

The challenge for ISRO's solid propulsion groups was to produce solid motors which not only gave the required performance but also operated reliably. They developed propellant combinations based on both imported and indigenous binders. There was considerable emphasis on ensuring that critical chemicals, such binders, were produced within the country.

The Propellant Fuel Complex was established to scale up laboratory processes and also undertake the production of binders and other chemicals. The Rocket Propellant Plant

(RPP), which cast the Centaure motors, was expanded and improved to meet the SLV-3 requirements. The Solid Propellant Space Booster Plant (SPROB) was set up at Sriharikota. SPROB was designed to cater to the needs of the SLV-3 as well as future requirements for bigger launchers. The Static Test and Evaluation Complex (STEX) where the solid motors could be tested also came up at Sriharikota. Steps were initiated for all four of these facilities in 1969-70, even before the SLV-3 project received the government's formal approval.

The motor cases of the SLV-3's third and fourth stages as well as its heat shield would be of fibreglass. In addition, the nozzles of the solid motors too used heat-resistant cloth to protect them from the high-temperature gases. As a result, work on composites also accelerated and the Reinforced Plastic Centre (REPLACE) was set up in Trivandrum.

The SLV-3 solid stages

The four solid stages of the SLV-3 closely resemble their Scout counterparts in dimensions, weight and the sort of material used for motor casing. There was, however, an important difference when it came to the SLV-3 first stage. Unlike the monolithic single-grain configuration used in the Scout, the SLV-3 first stage was made up of three segments. There was good reason for this change. The monolithic configuration required a single grain of 8 tonnes to be cast. Such a grain could be cast only after SPROB, the new solid propellant facility at Sriharikota, was successfully commissioned. Any delays in the commissioning of SPROB would then have a direct impact on the SLV-3 schedules. With a segmented configuration, on the other hand, the RPP in Trivandrum could be easily upgraded to produce the first stage segments in time. Each of these was 1 metre in diameter, about 3 metres long and weighed around 3 tonnes.

Segmentation is not without its difficulties. As the disaster of the space shuttle Challenger demonstrated, the slightest weakness in the joints between the two segments can be catastrophic. It is remarkable how smoothly the development of the SLV-3 first stage went, considering that ISRO had never made such a large motor — it had a diameter of one metre — was using a new binder system, and had no prior experience with segmented motors. The segmentation technology was first tested and proven in 300 mm diameter motors. A year later, during 1975-76, the first full-size segmented first-stage motor was ground-tested.

Since the first stage was segmented, the second stage, carrying 3.15 tonnes of propellant, became the largest single solid propellant grain successfully cast and flown by ISRO till the PSLV was launched in 1993. Both the lower stages used steel, a type called 15 CDV 6 by the French, for the outer casing. This steel had been used for fabricating the casing of the Centaure sounding rocket. Using 15 CDV 6 made it possible to utilize the experience already gained in fabricating rocket casing out of it.

The upper two stages, like the equivalent Scout stages, had fibreglass casing to save weight. The heat shield too was of fibreglass, as in the Scout. The development of fibreglass and other composite components will be given in more detail later on in this chapter.

The fourth stage of the SLV-3 raises some interesting issues. In *Wings of Fire*, Kalam narrates the tale of how Sarabhai had brought Hubert Curien, president of the French Space Agency, CNES, to Trivandrum. According to Kalam, Curien was so impressed by what he saw that he asked whether the Indians could develop the solid upper stage for France's improved Diamant launcher.

But did the idea come from the French or was such a development suggested by the Indian side? Irrespective of who made the suggestion, it seems improbable that the French would have had any genuine interest in subcontracting the development of a Diamant stage to ISRO. In 1971 (the year Sarabhai died), Indian capability in solid propulsion was still limited, even in sounding rockets. The SLV-3's detailed design was completed only in 1972 and development began in 1973. So Indian capability to make stages for a launch vehicle had yet to be demonstrated in any fashion, even in ground tests. It is not very likely that the French, or indeed anyone else, would contract out the development of an entire stage of a launch vehicle to novices.

A French delegation did visit Trivandrum in January 1971 for discussions on the sounding rocket programme, and possibly to assess the level of Indian technology. Since Curien was not part of this delegation, one has to assume that his trip, when the issue of the solid stage for the Diamant launcher supposedly came up, occurred earlier. Anyhow, at the meeting of the ISRO-CNES Working Group on Launch Vehicles in Paris in October 1973, the French made it quite clear that there was 'limited chance' of the SLV-3 fourth stage being used for the Diamant. They bluntly stated that India still had a long way to go in solid propulsion.

Around this time, while the issue was at least being discussed by the two countries, the French reportedly tried to sell India solid propellant technology. Sarabhai apparently retorted that India could develop the technology on its own, quite a remarkable statement considering that India had imported French technology for the Centaure just a few years earlier. To prove the point, two solid propellant formulations, given the names Veli-21 and Veli-22, were developed. Veli-21 was made with imported carboxyl terminated polybutadiene (CTPB) resin and Veli-22 made with polypropylene glycol which was readily available in India because of its use by the plastics and foam industry. Some twenty four 2 kg test motors were made with each of the two propellant formulations. While Sarabhai reviewed the work inside an office room, these motors were ground-tested one after another outside, remembers V.N. Krishnamurthy, who was present and has since retired from the space programme.

The fourth stage of SLV-3 did, however, find use outside the SLV-3 project. The European Space Agency had offered to carry, free of cost, a small Indian satellite during the initial experimental launches of the Ariane launch vehicle. India gratefully accepted the offer and began developing APPLE, the Ariane Passenger Payload Experiment, a 600 kg experimental communications satellite. Ariane, like most launch vehicles, leaves a satellite in an elliptical orbit termed the geostationary transfer orbit (GTO). The satellite has then to use an onboard motor to move to the geostationary orbit, some 36,000 km above the equator, where it will match the earth's rotation and therefore appear stationary from the ground. This role of moving the APPLE satellite from GTO to geostationary orbit fell to a modified SLV-3 fourth stage. On 21 June 1981, a year after the first successful launch of the SLV-3, the APPLE's Apogee Boost Motor (ABM), as it was called, was fired and performed its task flawlessly.

As with the casing, the upper stages and the lower ones used different propellant resins. The first and second stages used imported polybutadiene-acrylic acid-acrylonitrile (PBAN) and the upper stages an entirely indigenous resin called HEF-20. When I spoke to various people involved in solid propulsion development during that period, the explanation I got was that such duality had been noticed when they looked at solid stages developed by other countries. PBAN was being used for the lower stages and CTPB for the upper stages. CTPB's better energetics and mechanical properties made it well-suited for the upper stage propellant.

If this were so, surely it made sense to use CTPB for both upper and lower stages. The reason I was given for CTPB not being used for the lower stages as well was that it was costlier than PBAN. Since the lower stages were much bigger and needed more resin, these stages would become more expensive to produce if CTPB were used. HEF-20, the indigenously developed functional equivalent of CTPB, also worked out costlier than PBAN.

While such reasoning provided the justification, the internal politics of VSSC may also have played a part in the choice of propellants. Development of propellant formulations and casting of the lower two stages was carried out at RPP under Kurup. The upper two stages were similarly handled by Gowariker's team at SPROB and HEF-20 was developed by his Propellant Engineering Division. The strained relations between the two men may have been an additional reason for not using HEF-20 in the lower stages.

Setting up facilities for solid propulsion

With a project like SLV-3 on hand, ISRO had to use the facilities and capabilities it had built up. Kurup and his team at the RPP, who had established the Centaure capability, were assigned the responsibility for production of the lower stages of SLV-3. This was no small task. This would be the first time they'd be dealing with PBAN, an advanced propellant binder. PBAN is still extensively used in the United States, including in the world's largest solid motors, the space shuttle's Reusable Solid Rocket Boosters. While the upper stages of the SLV-3 would also be using high-energy case-bonded propellants (though based on a different resin), the two lower stages were much larger and heavier. As explained earlier in this chapter, difficulties of making solid propellant motors increase with their size.

RPP was intended to make the Centaure rockets. The Centaure's lower stage was just 300 mm in diameter and contained 232 kg of propellant. The RPP group was able to utilize the same facilities to cast a 700 kg propellant grain with a diameter of 560 mm which would later form the lower stage of the RH-560, the largest sounding rocket ever built by ISRO. But the SLV-3 lower stages were much bigger than these motors. Each of the three segments of the first stage were 1 metre in diameter and contained close to 3 tonnes of propellant. The second stage had a diameter of 0.8 metre and carried 3.15 tonnes of propellant. So RPP had to be expanded to cope with the larger size of the SLV-3 motors. Facilities to mix and cast larger quantities of propellant were needed. Ovens were required where these big motors could be 'cured'. The expansion of RPP to meet these needs was initiated in 1969, a year before the SLV-3 project was approved.

The prime responsibility of Gowariker's Propellant Engineering Division was to indigenously develop propellant resins and other chemicals. Gowariker was a hard taskmaster, continually driving his team to their limits. But he was also respected by them. As several people who worked under him in those days testify, his emphasis on self-reliance was a key factor in ensuring that ISRO did not remain dependent on imports for vital chemicals used in solid propulsion. The American embargoes, when they came, were powerless to affect this critical part of the launch vehicle programme.

Even before they had a laboratory scale process ready, the Propellant Engineering Division approached Sarabhai with a proposal to establish the Propellant Fuel Complex (PFC) at Thumba. Its role, as Gowariker apparently explained to Sarabhai, was twofold. A chemical process shown to work in the laboratory had to be demonstrated as feasible on a much larger scale before industrial production could be attempted. The PFC could be used for such scaling up studies. At the time, there was no other pilot plant facility in the country for polymer production. Gowariker envisaged that the PFC could also be used for production purposes. Both these roles have made sure that the PFC continues to be operational even today. Should the Indian company which produces HTPB (hydroxyl terminated polybutadiene), the resin used in all ISRO solid motors currently, be unable at any time to supply the resin, the PFC can make sure that the space programme is not halted.

Sarabhai, with his experience in managing chemical and pharmaceutical companies, was quick to see the reasoning and gave his approval for setting up the PFC. I was told that the foundation stone for the PFC was laid in February 1972, two months after Sarabhai's death, by Prof. M.G.K. Menon, then chairman of ISRO. The PFC became operational two years later.

It was again Gowariker who suggested establishing the Solid Propellant Space Booster Plant (SPROB) at Sriharikota and got the go-ahead from Sarabhai. Quotations were obtained from companies in the United States and France for SPROB. Their proposals provided an understanding of the precautions which needed to be taken, the various types of equipment required and their function, as well as listing the suppliers of these equipment, according to Gowariker. A team then spent ten intensive days, touring facilities in France and the United States. The Indians made extensive notes of what they saw and learnt. 'The trip was an eye-opener,' says Gowariker.

The foreign companies had demanded Rs 50 crore, half of it in foreign exchange, says Gowariker. His group decided that they could set up the facility for just Rs 8 crore, with Rs 80 lakh in foreign exchange. The SPROB proposal was formally cleared in 1971, with a sanctioned cost of Rs 7.92 crore. The plant would be capable of casting solid propellant grains up to 2 metres in diameter and weighing 10 tonnes. The 10 tonnes capability meant that it could cast the whole SLV-3 first stage as a single grain, rather than in segments.

But what is truly amazing is that SPROB's production capacity was fixed at 500 tonnes of propellant a year. That capacity was enough to produce well over 30 SLV-3s every year! It was only with the PSLV, which flew two decades later, that such a capacity was needed. As a result, when funding for the whole space programme became tight, the SPROB capacity was halved, retaining provisions for its later expansion. Even then, by the time SPROB was commissioned in March 1977, it cost Rs 7.74 crore.

Much of the equipment for SPROB was made within the country, although some crucial equipment, such as large mixers and the linear accelerator needed for X-raying the finished propellant grains, had to be imported. A remote-controlled trimming machine, for instance, was designed and fabricated with the help of the Central Machine Tools Institute at Bangalore.

An ISRO document remarked that 'with the commissioning of SPROB, India has in terms of monolithic propellant grain size, one of the largest solid propellant manufacturing capabilities in the world'. The *Annual Report* observed that 'SPROB was designed and built with completely indigenous know-how and skills developed at Vikram Sarabhai Space Centre, Trivandrum. It ranks as a major achievement of Indian technology, engineering skills and practice'.

After their natural-rubber based propellant which flew in a few RH-75 sounding rockets, Muthunayagam's Propulsion Division (often shortened to PSN) did not undertake development of any more solid propellant formulations. Once Kurup's and Gowariker's groups developed the formulations and established their characteristics, the data would be passed on to the propulsion group. They would then design the motor, including the shape of the central port of the propellant grain. Since the area of the propellant exposed to burning determined how much gas was produced and hence the thrust generated, the shape of the port was very important. Muthunayagam also had the responsibility for development of liquid systems for the SLV-3, which is dealt with in the next chapter.

The establishment of the Static Test and Evaluation Complex (STEX) for testing large solid motors was an idea put forward by Muthunayagam and approved by Sarabhai. Sriharikota was the logical place to put up STEX from the point of view of safety, space for expansion as well as proximity to SPROB, where the motors would be cast.

For STEX, Muthunayagam visited various centres in the United States, looking for information on the sort of test facilities needed as well as help in setting them up. At the time, ISRO did not know enough even to define its requirements for test facilities, according to him. As narrated earlier, at Sarabhai's request, Dhawan made arrangements for both of them to visit the Jet Propulsion Laboratory's test facilities. But the United States took a tough stand on any sort of help in establishing solid motor test facilities in India.

The Indians then approached the French space agency, CNES. 'By that time, we had a reasonable idea of what was needed,' says Muthunayagam. Through CNES, technical cooperation with the French company, Société Europêene de Propulsion (now Snecma Moteurs' rocket engine division), was arranged. A SEP newsletter of 1973 reported that the company had recently signed a technical cooperation agreement with ISRO. The agreement would help 'the Indian specialists avail themselves of SEP experience in the field of solid propellant rocket engine testing, and to give them advice in the development of a test centre for such solid propellant engines, that will be located at Sriharikota on the eastern coast of India', it added.

After the cooperation agreement was signed, an ISRO team went to France. There, with SEP's help, they designed the various test facilities, including the critical six-component test stand. Apart from giving India access to the technology for the test facilities it needed, the cooperation arrangement with SEP had another impact. It prepared the ground for ISRO and SEP to sign another, much more important agreement a year later. Under this agreement, Indian engineers would work with the French on the development of their Viking liquid engine and acquire the technology for it. The deal became ISRO's most important technology acquisition and was achieved at a very low price.

But SEP did not have experience with high altitude test (HAT) facilities which could simulate near vacuum conditions. The HAT was essential for fully testing the fourth stage of the SLV-3, which would also be flying as the apogee boost motor to take the APPLE from the transfer orbit to geostationary orbit. ISRO was, however, able to get the help of the German space agency, DFVLR, in designing the HAT facility.

Most of the fabrication needed for STEX was done within India, although critical parts and machines had to be imported. When an American company was prevented from supplying a 16 tonne vibration table, a British company stepped in. The latter was only making 2.5 tonne vibration tables at the time and developed their first 16 tonne table for ISRO. STEX too was designed with the need for later expansion kept in mind.

Although the STEX proposal was formally cleared in 1970, work on the first phase to establish facilities for sea-level testing began only in 1972 and took four years to complete. The HAT facilities took a couple of years more. In the end, STEX cost close to Rs 6 crore. As with SPROB, STEX too would be expanded to meet PSLV requirements. After STEX became available, the testing of large solid rocket motors at Trivandrum stopped and all such testing has since been carried only at Sriharikota.

Development of indigenous resins

The Propellant Engineering Division headed by Gowariker began development of a propellant for the upper stages with the CTPB resin. CTPB, for reasons explained earlier, was considered more suitable for upper-stage applications. Although small quantities were supplied for trials, the American company involved refused to accept orders for bulk supply. 'This is a typical American technique to delay development,' remarks K. Sitaram Sastri, currently VSSC's deputy director for Polymers, Chemicals and Materials.

The decision was then taken to develop CTPB indigenously. A key raw material needed was butadiene gas, then produced only by the Synthetics and Chemicals Company situated at Bareilly in Uttar Pradesh, in order to make butadiene rubber. Specially made cylinders, each carrying 20 kg of the gas and later 100 kg, were used to ship the gas by train to Trivandrum. But problems are said to have cropped up developing a process which would consistently yield CTPB of suitable quality. Bringing the gas in cylinders by train was cumbersome and time-consuming. Time was something they didn't have, as delays in developing the resin would affect stage development.

The group at the Propellant Engineering Division decided to simultaneously try a different approach. This was to use polybutadiene rubber which, unlike butadiene gas, could be transported to Trivandrum in bulk without any difficulty. Using polybutadiene rubber as the starting material, instead of butadiene gas, is not an idea which appeals to a scientist. Much energy and raw materials have to be put in to produce the rubber from butadiene gas. Since the rubber has to be broken down first, more energy and raw materials have to be expended before it can be processed further. It is therefore a less efficient and more expensive method than if the gas were the starting material. But the process based on the rubber worked well and yielded CTPB of excellent quality. Later analysis revealed that its chemical structure was slightly different from that of CTPB and it was termed lactone-terminated polybutadiene (LTPB). In ISRO, it was called HEF-20 (HEF standing for high energy fuel). HEF-20, the functional equivalent of CTPB, is said to have been five times costlier than imported PBAN.

The HEF-20 process was reportedly proven at the laboratory scale by mid-1971. Since work on the Propellant Fuel Complex (PFC) had started only a few months earlier, there was no pilot plant where the process could be scaled up. A house was hired in the Karamanai area of Trivandrum and a makeshift pilot plant fashioned from steel vessels and other off-the-shelf materials. Large-scale production of HEF-20 began only after PFC became operational in 1974.

By 1972-73, HEF-20 was substituted for imported CTPB in the Veli-21 propellant formulations and subscale motor tests were carried out. The *1975-76 Annual Report* of the Department of Space says that HEF-20 had been successfully substituted in propellant formulations for the SLV-3 third and fourth stages. HEF-20 was also used in the Apogee Boost Motor of the APPLE satellite.

The Propellant Engineering Division later developed a process to make PBAN indigenously. The *1975-76 Annual Report* speaks about facilities for PBAN production being commissioned. A few years later, the PBAN production capacity was augmented to 7.5 tonnes annually. By 1980, the possibility of having the public sector petrochemical company IPCL produce PBAN at its Baroda plant was being explored.

But despite all this activity, indigenously produced PBAN never flew on any Indian launch vehicle. Even though the United States, as ISRO had been expecting all along, embargoed PBAN exports for a period after the SLV-3 launch, there were still enough stocks of imported

PBAN for the remaining SLV-3 flights and even for the first two flights of the Augmented Satellite Launch Vehicle (ASLV).

By then ISRO had successfully indigenized the production of HTPB, a better binder than PBAN. India recognized HTPB's potential early and chose it for the PSLV solid propellant. ISRO's interest in PBAN died out and it concentrated all its efforts on HTPB. Once HTPB became available, all ISRO solid motors used only that resin. Even in the ASLV, PBAN was substituted with HTPB for the third and fourth launches in 1992 and 1994

Some of the Propellant Engineering Division's most controversial projects, which still make many people in ISRO get hot under the collar, arose from its work on castor oil. PED began working on castor oil and its principle derivative, 12-hydroxy stearic acid (THSA), while developing a chemical needed in solid propellant manufacture. It then began to explore other uses for THSA, exploiting its chemical structure. The division tried using castor oil itself as a propellant binder. It could be done, but its properties varied too much to be useful in that role. The division then undertook a catalytic conversion of THSA into a polymer form called a polyol. The result was ISRO polyol, an entirely new low-cost, high-performance propellant resin.

Propellant formulations made with ISRO polyol appeared to compare quite favourably with those involving other high-performance propellant resins. ISRO polyol had the advantage that it was much cheaper than resins derived from petrochemicals. The chemical structure of ISRO polyol made a separate plasticizer (added to increase processability of the propellant) unnecessary in many propellant formulations and also made it less vulnerable to the effects of ageing. The major limitation of ISRO polyol was that there was too much variability in the properties of the propellant formulations made with it. This may have had to do with variation in the THSA produced or the presence of some stearic acid.

The *1976-77 Annual Report* of the Department of Space remarked that the special features of ISRO polyol-based propellants were their low cost, negligible import content and suitability for casting large motors. 'This propellant may be used for casting the monolithic version of the SLV-3 first stage,' it added. In a paper presented at a professional conference abroad in 1976, Gowariker and his team said that ISRO polyol-based propellants 'have demonstrated the feasibility of their use in 10 tonne monolithic sizes'.

This was not just talk. A monolithic SLV-3 first stage using an ISRO polyol-based propellant formulation was indeed cast at SPROB. The facilities at SPROB had, after all, been built keeping in mind the requirement that motors up to 2 metres in diameter and solid propellant grains up to 10 tonnes would be cast here. The test of the monolithic motor was carried out on 27 March 1977, the day SPROB was commissioned. The motor exploded. Dhawan reportedly walked out. It is unlikely that the failure as such irritated him. Failure, after all, is a common occurrence in space-related trials. From all one hears, it was probably the way in which the monolithic development and test were carried out which met with his disapproval. The monolithic motor, though technically a sound idea, seems to have been pushed too hard and too fast so that the test could coincide with the commissioning of SPROB.

Although the problem which caused the failure seems to have been readily solvable, the test was not repeated. The reasoning is not hard to understand. The first launch of the SLV-3 was less than two years away. The segmented first stage had already been successfully tested and there was no pressing need for a monolithic motor. The last thing the SLV-3 project needed at this point was a tussle between Kurup and Gowariker over whether the first stage should be the segmented PBAN type or the monolithic ISRO polyol-based version.

Almost two decades were to pass before a monolithic 1 metre diameter 10 tonne motor was tested in November 1994, this time using an HTPB-based propellant. Although the PSLV uses six motors of this size attached to its first stage, the segmented construction perfected during the SLV-3 project is still preferred so that the facilities at the Rocket Propellant Plant can be fully utilized.

ISRO polyol was never used in any ISRO launch vehicle. It was, however, substituted for polypropylene glycol in the RH-300 propellant. The RH-300 with IPP-40, a propellant formulation with ISRO polyol, was successfully flown in January 1983. A couple of years later, the same formulation was again used in the RH-300 Mark-II, a single-stage sounding rocket to replace the two-stage Centaure.

In the mid-Seventies, the Propellant Engineering Division became involved in a highly publicized and, within ISRO, highly controversial effort to use their castor oil-related knowledge for helping society at large. According to what I was told, the division got interested in the non-space possibilities of their work after the Khadi and Village Industries Commission (KVIC) approached ISRO for help in finding new uses for non-edible oilseeds to help generate rural jobs and income. What the division came up with was 'space crude'.

Conversion of vegetable oils into petroleum products was not a new discovery. Germany used this method in

World War II to keep its war machinery going after its crude oil supplies were blocked. The division's achievement was its catalyst, which made low-pressure and efficient conversion possible. The process worked with a wide variety of non-edible oilseeds found commonly in India.

As it happened, space crude came on the scene when India was reeling from the effects of an oil price hike enforced by the Organization of Petroleum Exporting Countries (OPEC). The Indian government had been forced to sharply raise the administered prices for petrol and other petroleum products, never a popular measure. There was also a serious foreign exchange shortage. Here, it seemed, was a technology, developed in the course of India's efforts to build space capability, which could turn millions of tonnes of oilseeds to much-needed petroleum and create a large number of jobs in the process. A renewable and plentifully available resource would be used, a need would be met and foreign exchange be saved. To use modern day phraseology, it seemed the ultimate win-win situation.

When Dhawan delivered the Dr A.L. Mudaliar Memorial Lecture at the Indian Institute of Technology, Madras, in February 1976, his talk was devoted to space Crude. He explained that the ISRO process converted the non-edible oils into a liquid which accounted for 65 per cent of the feedstock, as well as a gaseous fraction accounting for about 23 per cent. The liquid fraction could be used to produce petrol, kerosene, diesel and lube oils. The gaseous fraction provided olefins, the starting point for cooking gas and many petrochemicals. In addition, the oilcake and hulls of the seeds could be used as organic manure or converted to furfural, a chemical needed by the plastics and dye industries.

Dhawan observed that 7 million tonnes of non-edible oilseeds were available all over the country each year, the bulk of which came from uncultivated plants and usually went waste. If 200 units to process these oilseeds were set up, they could provide direct employment to nearly a million people, three-quarters of them unskilled and semi-skilled. Plants could be put up with a capital investment of Rs 190 crore and operating costs of Rs 461 crore. The value of their products, before profit and taxes, would be Rs 487 crore annually. With 150 per cent tax ex-factory, the revenue to the government worked out to about Rs 730 crores a year, Dhawan pointed out.

Experimental oilseed collection studies were mounted in Bihar and Andhra Pradesh. But that is the last one hears about space crude in the annual reports. Two factors appear to have made the project unviable. Some non-edible oilseeds already had a higher value product for which they were being used — soap-making, for instance. The cost of collecting seeds which were not being commercially exploited was too high, reportedly because of the role played by middlemen, to make petroleum production economically viable.

Space crude is still a controversial issue in ISRO circles. Some have had reservations on technical grounds. Many more were convinced that it was oversold as part of Gowariker's empire-building and self-promotion. Should ISRO have pursued a venture which was outside its principal mandate of developing space technology? S. Chandrashekar, who was on the ISRO headquarters staff at the time and is now on the faculty of the Indian Institute of Management, Bangalore, believes the organization had done the right thing by exploring what appeared to be a promising and socially relevant technology. The strength of ISRO lay in that mere promise was not allowed to go out of hand. When trials showed that the project was unviable, it was also stopped. The money which ISRO spent in the process was small and the technical capability and confidence built up as a result quite considerable, says Chandrashekar.

ISRO polyol did, however, go into commercial production for a while. As in solid propellants, ISRO polyol could be a cheaper substitute for polypropylene glycol used in the foam industry as well as for making insulation and other applications. ISRO tested rigid and flexible foam formulations based on its polyol. Some test marketing was carried out by the Kerala State Industrial Development Corporation. The technology for ISRO polyol was transferred in 1980 to a joint sector company, Malabar Polyols and Allied Products. ISRO provided technical assistance for setting up and commissioning the 1,000 tonnes per annum plant at Kuttiupuram in north Kerala. The plant went into full production in 1987. But after some years of production, the company reportedly went sick as a result of various managerial problems. The technology was later also given to United Breweries in Bangalore.

All said and done, ISRO's work on polyols was new and innovative. The clearest indication of this is that ISRO took out patents for certain polyols as well as for the process to make them, not just in India but in the United States and Britain as well.

Ammonium perchlorate

Ammonium perchlorate is the largest single ingredient in a composite propellant, responsible for about 70 per cent of its total weight. Since the total solid propellant in SLV-3 was 13 tonnes, the ammonium perchlorate needed for a single launch would be around 9 tonnes. Four launches would require 36 tonnes of it. Since each motor would have to be tested on the ground many times, the amount of ammonium perchlorate needed for the entire SLV-3 programme would be much more than 36 tonnes. As ISRO built larger solid motors for its future launch vehicles, the demand for this ingredient would sharply increase.

Initially, the ammonium perchlorate was imported from France as part of the Centaure programme. But ISRO was well aware that depending on imports for such a strategic material would make its programme vulnerable to foreign embargoes. The very first annual report of the newly created Department of Space stated that 'since ammonium perchlorate is an important constituent of composite propellants and is in short supply in the country, it is planned to set up a plant for its manufacture'.

ISRO, in fact, kept two strings to its bow. The match manufacturer, Wimco, agreed to produce ammonium perchlorate for the programme. Wimco is said to have begun supplying ammonium perchlorate to ISRO from the mid-Seventies and to have continued to do so till the early Eighties. ISRO did not, however, want to become wholly dependent on one supplier, especially since Wimco had foreign equity holding.

It decided to establish a plant of its own at Alwaye (now called Aluwa) in Kerala and the task fell to the Propellant Engineering Division. It opted to base the process at this plant on a new cost-reducing method developed at the Central Electrochemical Research Institute (CECRI) at Karaikudi in Tamil Nadu. One of the steps in the production of ammonium perchlorate involves electrolysis, the passing of current through a solution. Conventionally, one of the electrodes would be made of platinum, a metal more expensive than gold. CECRI found that the platinum electrode could be substituted with one of lead dioxide deposited on a graphite substrate. Lead dioxide by itself was too fragile to be used as an electrode.

But the CECRI's process was essentially a laboratory scale method. In an industrial production environment, the graphite substrate electrodes gave trouble. Although these electrodes were supposed to last a year, the lead dioxide coating would often crack and peel off much earlier. Once the coating broke away, the graphite substrate would be rapidly corroded, contaminating the solution. As a result, the product quality varied from batch to batch. The problem turned out to be impurities in the materials used to prepare the electrodes. ISRO scientists resorted to purifying the raw materials themselves and were able to extend the life of electrodes.

In the mid-Eighties, ISRO discovered a better solution. It substituted titanium for the graphite substrate. Although titanium was more expensive than graphite, it compensated with much longer life. It had the added advantage of being immune to corrosion even if the lead oxide coating broke. But lead dioxide could not directly be deposited on titanium as a layer formed which inhibited the flow of current. A technology had to be developed of depositing other metal oxide layers on top of titanium before the lead oxide layer was added, says Sreenivasa Setty, who was closely involved in establishing the ammonium perchlorate production. Setty later became scientific secretary of ISRO.

The Ammonium Perchlorate Experimental Plant (APEP) was commissioned in 1978 and formally inaugurated by the chief minister of Kerala in 1979. But during the initial years, the plant produced only about 40 tonnes of ammonium perchlorate a year, against its capacity of 150 tonnes. One reason was that the demand from the launch vehicle programme had not picked up (Wimco too, it must be remembered, was supplying ammonium perchlorate at this time). But the electrode problem also appears to have contributed to the low production levels; one annual report spoke of 'debottlenecking' the plant so that it could reach its rated capacity. Once the titanium substrate electrodes were in place, production rapidly climbed. During 1984-85, APEP produced 160 tonnes of ammonium perchlorate, in 1989-90 it produced 200 tonnes. Not only did ISRO have no shortage, but it soon had an ammonium perchlorate mountain on its hands. One tonne of the chemical was sold to the Indonesian Space Agency in June 1992.

The lower two stages of SLV-3 reportedly used ammonium perchlorate from Wimco while the upper two stages utilized the perchlorate produced at the Alwaye plant. If true, then the Kurup-Gowariker rivalry may once again have been the reason. The Alwaye plant was set up by Gowariker's group. By the mid-Eighties, however, Wimco stopped supplying ammonium perchlorate and ISRO depended entirely on APEP. The plant came into its own with the Polar Satellite Launch Vehicle programme since each PSLV needs nearly 150 tonnes of ammonium perchlorate.

Composites

ISRO copied the Scout model in its use of fibreglass for the third and fourth stage motor cases as well other composites for the SLV-3. Despite limited prior experience, they were nevertheless able to develop some twenty-seven different components of the quality required for the SLV-3 and have them ready within the stipulated time.

In any composite, a fibre or some other reinforcing element is embedded in a matrix and the combination gives the material its distinctive properties. In fibre reinforced plastics (FRP), the fibres, either as strands or as woven material, are embedded in a plastic resin which hardens when it sets. Alternatively, the filaments can be wound and held in place by the resin to make things like motor cases.

The fibre and the resin can be varied to tailor the material to meet specific needs. Low weight, high strength, and good insulation from heat are some of the properties usually associated with FRPs. The fibreglass motor cases of the SLV-3, for instance, needed to have the required strength without weighing as much as the steel used for the lower stages. The heat shield of the SLV-3 had to keep the satellite below 40 degrees Celsius even though the outside of the launch vehicle would get very hot as it raced through the dense atmosphere. The ablative linings at the exit of the solid motor too were made of composites. These ablative linings would take the brunt of the hot gases and gradually burn away, protecting the metal or composite material behind.

The credit for initiating work on composites goes to Abdul Kalam. He was responsible for the payload development for sounding rockets and the initial efforts in composites were to make some non-metallic components needed for this purpose. These first ventures used handmade moulds, remembers C.R. Sathya, who headed the FRP effort and is now with Tata Advanced Materials. Put into the oven for curing, they came out charred!

The first filament winding machine built was hand-cranked and used coir ropes to drive wheels and pulleys. It

Abdul Kalam, is on Sarabhai's right and C.R. Sathya on his left.

was very crude and funny looking, although it did serve to prove the concept. When Sarabhai came to Trivandrum on one of his periodic visits, Kalam brought him over to see the machine. While others who came along with Sarabhai laughed heartily when they saw this primitive-looking contraption, Sarabhai's reaction was markedly different. Could a motor-driven machine be got ready in three months time when the Prime Minister, Indira Gandhi, would be coming to dedicate the Thumba Equatorial Rocket Launching Station, he asked.

A motorized and more sophisticated version of the filament-winding machine, which used an old gearbox from a car, was indeed formally switched on by the Prime Minister. The machine is said to be in use even today to make cases for solid motor igniters. The composites group would, in fact, continue this tradition of making the machines they needed. Only few of their equipment, such as the autoclave, would be imported.

The composite requirements of the SLV-3 were demanding. Fortunately, two people, one of whom was Sathya, had been sent in mid-1971 to Sud Aviation in France for three months' training in making the composite nozzle for the Centaure. Sathya then spent another three months visiting various establishments in Europe and the United States which were working on composites, including those making machinery for it. He visited NASA's Wallops Island facility and saw the Scout motors with composite casing. He returned to Trivandrum with ideas of how ISRO could go about meeting the SLV-3 requirements as well as future needs.

The Fibre Reinforced Plastics (FRP) Division was created in 1971 and Kalam also proposed the setting up of a Reinforced Plastics Centre (REPLACE). Although it was originally intended as an industrial unit to meet the requirements of the SLV-3 programme and of other space projects, financial constraints led to it being curtailed to meeting the SLV-3 needs alone. REPLACE was commissioned in November 1976.

The glass fibres were purchased from Fibreglass Pilkington in Bombay and the plastic resin from Ciba, also in Bombay. Although high silica glass cloth needed for ablative linings had initially to be imported, by 1977-78 a process to increase the silica content in commercially available glass cloth was ready. Similarly, although the resin for ablative purposes had at first to be imported from France, this resin too was subsequently indigenized by the Propellant Engineering Division.

By 1975-76, SLV third and fourth stage fibreglass motor cases were ready for testing. Two full-size fibreglass heat shields were also made. None of the composite parts gave any problems in the four flights of the SLV-3. The modern non-destructive testing facilities, based on X-rays, ultrasound and other methods, which had been established were an important element in this success, says Sathya.

When the fourth and last flight of SLV-3 lifted off in April 1983, the launch vehicle's fourth stage had a new motor case. Instead of fibreglass, it used a lighter and higher strength fibre known as Kevlar which entered commercial use only in the early Seventies. Kevlar is actually the trade name given by DuPont, the American chemical company which created it. Since a Kevlar motor case can be thinner than a fibreglass one without compromising on strength, the weight of the casing can come down by as much as 30 per cent. Since the motor was also slightly longer, 45 kg of additional propellant could be loaded.

The Augmented Satellite Launch Vehicle would use an identical Kevlar fourth stage. The experience would be vital in developing the Polar Satellite Launch Vehicle's third stage, one of the largest upper stage solid motors in the world whose casing too would be of Kevlar.

In conclusion

By successfully meeting the challenges posed by the SLV-3, ISRO established a top-class team in solid propulsion, a team that was dynamic and confident of its abilities. The creation of such competence led, in turn, to major achievements during the development of the Polar Satellite Launch Vehicle (PSLV). In the process, ISRO's solid propulsion group established itself as having world-class capability.

ISRO was always aware that its launch vehicle programme would be regarded with suspicion by the West and that it had to prepare as quickly as possible for any embargoes which might come. The Propellant Engineering Division, led by Gowariker, played an important part in indigenizing many key chemicals, including the resins and ammonium perchlorate needed for solid propellants.

Dismissing the castor oil related work on ISRO polyol and space crude as failures is missing the wood for the trees. They may not have been successful ventures, but their development created competence in terms of a deeper understanding of the chemistry involved and improved the group's problem-solving skills. It created too an ethos of not dismissing ideas just because they were unconventional. It nurtured a culture of being unafraid of taking risks inherent in any technology development.

CHAPTER

5

Early Initiatives in Liquid Propulsion

FOR THE PRESENT at least, launch vehicles can be propelled only by two means – either by solid motors or by engines burning liquid propellants. Although a liquid engine system is mechanically more complicated than a solid motor, its greater efficiency and other advantages make it the preferred option for most launch vehicles. But solids got a head start in ISRO during the development of sounding rockets. This advantage got consolidated further when Sarabhai decided, for reasons explained earlier, to have the SLV-3 as an all-solid launch vehicle modelled on the Scout.

But even in the SLV-3, and the Scout, for that matter, small liquid engines, called thrusters, were needed. These thrusters served the vital function of controlling the launch vehicle's orientation. Although these thrusters were small, their development would not have been possible if some work on liquid engines had not already been done. In ISRO, development of liquid engines began later than work on solid engines. Those involved were aware both of the difficulties of liquid propulsion as well as its advantages for use in launch vehicles.

In contrast to the seeming simplicity of solid motors, liquid propellant stages are undoubtedly more complex. They need tanks to hold the propellant and a separate engine where it will burn. Plumbing is needed for the propellant to flow through to the engine. Valves and regulators have to control the flow. The combustion chamber and other parts of the engine may have to be protected from the hot gases generated during combustion. The powerful vibrations and violent jostling which a launcher is subjected to during its flight can set up disturbances in the propellant flow, with potentially serious consequences. A liquid engine and its stage require far more high-precision mechanical fabrication and assembly than a solid stage.

So why do rocket designers want liquid engines? For one thing, liquid propellants are more energy efficient. Producing 1 tonne of thrust would require about 3.8 kg of composite solid propellant being burnt every second. If liquid engines were used, less than 3.5 kg of propellant would be needed. With cryogenic fuels, involving liquid hydrogen and liquid oxygen, only a little over two kg would have to be burnt per second. Propellants account for three-quarters of any launcher's weight. Thus if the amount of propellant which needs to be carried can be reduced, the launcher can become smaller, lighter and probably cheaper too. The payload can then form a larger proportion of the weight at lift-off.

Liquid engines have a number of other advantages too. They can be tested on the ground before being flown. In some of them, their thrust can be varied. More important, they can be stopped and restarted, if necessary. For all these reasons, launchers the world over have tended to rely on liquid engines for their core propulsion, with solids providing extra thrust at lift-off.

As we shall see, although liquid propulsion was a late starter in ISRO when compared to solids, it has steadily grown in importance. Liquid propulsion began in a small way in Muthunayagam's Propulsion Division. At the time, much of the activity of this division related to the design and development of solid motors. By April 1980, liquid propulsion had grown sufficiently for a separate Auxillary Propulsion Systems Unit (APSU) to be set up. In 1985, the Liquid Propulsion Systems Unit (LPSU) was created to consolidate all liquid propulsion activities. A couple of years later, this unit became a full-fledged ISRO centre, the Liquid Propulsion Systems Centre (LPSC). By the time Muthunayagam left the directorship of LPSC in 1995, liquid propulsion had grown to employ some 1,500 people.

ISRO's first liquid engines

The Robinson Crusoe-ish approach of improvising with whatever was available was very much in evidence during the early days of liquid propulsion in ISRO. Much of the early work in this area was based around two readily available propellants — aniline for fuel and red fuming nitric acid (RFNA) as oxidizer. The space people apparently cadged the nitric acid from the Air Force who were using it for their Soviet-made missiles. 'The Air Force gave us the nitric acid from their old stocks and the nitric acid we got was neither red nor fuming,' one person recalled. Aniline was widely used in the the dye industry and so was commercially available. But aniline is also very carcinogenic, although this aspect was not known at the time. Protective clothing for handling these two dangerous propellants consisted in those days of just rubber boots and rubber gloves!

RFNA and aniline are storable at room temperature and are also hypergolic, which means that they burn when they come into contact with one another. In one early experiment to demonstrate their hypergolic behaviour, reportedly done in front of Sarabhai, the propellants were poured into separate burettes mounted on laboratory stands so that their streams could mingle. The burettes' taps were then opened. But, in order to have proper combustion, these propellants have to be mingled in fine droplets, not poured down in a stream. They also have to be mixed in the right proportions. So instead of the propellants burning, the experiment resulted in the grass and other plants in the neighbourhood getting scorched by the powerful acid.

The next experiment was an improvement. The Liquid Propellant Motor, given the designation LPM-0, was developed and demonstrated to Sarabhai around late 1968. It was still a fairly amateurish effort. It consisted basically of a metal chamber 3 to 4 inches long and perhaps one and a half inches in diameter. It could be comfortably carried in one hand. The propellant tanks were stainless steel tubes closed at both ends. Pressurized gas would enter at the top of the tank and push the propellant out at the bottom, through a single injector, and into the combustion chamber. There were no valves and so the only way to control the flow was by opening and closing the tap at the gas cylinder

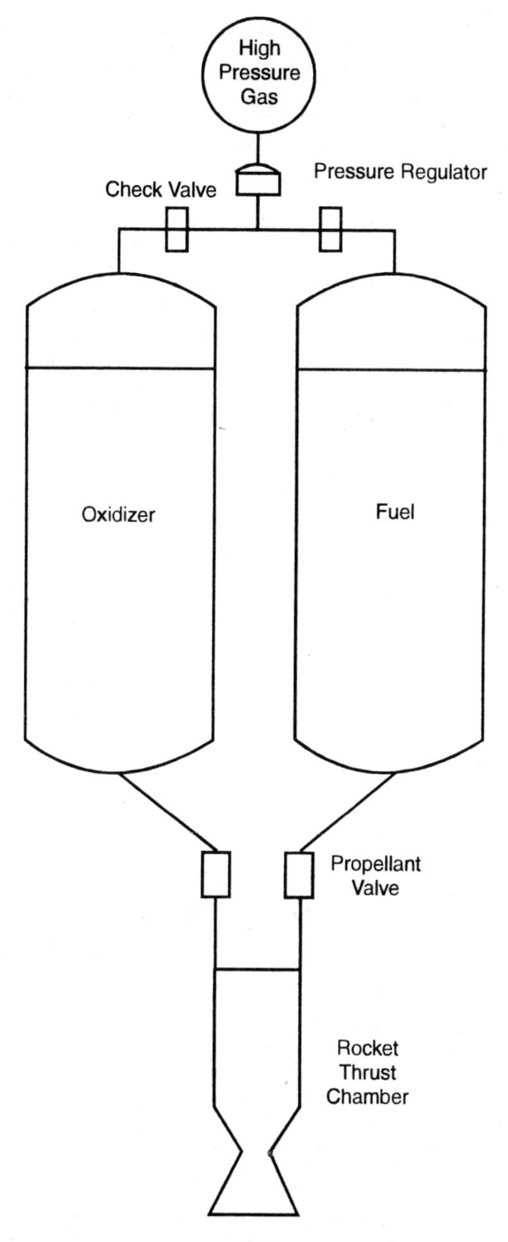

Schematic of a Pressure-fed Liquid Propulsion System

itself. Since there was still no method of controlling the mixture ratio, instead of combustion there would often be propellants streaming down from the engine. Once in a while, conditions would be just right and combustion would occur in spurts.

An engine developing 200 kg thrust was then built. This one had multiple injectors and various injector configurations were experimented with. The combustion chamber's walls were protected from the hot gases with an ablative lining. An ablative protects the material behind by absorbing the heat and gradually burning away. The ablative used in this case is said to have been asbestos filled with phenolic resin. As the quality of the ablative lining improved, the duration for which the engine could be fired increased from 10 seconds to about 45 seconds.

Rohini RH-125 sounding rocket motor cases, which had been discarded after ground tests, were cleaned and made to serve as propellant tanks. Joints which showed a tendency to leak were fixed with Teflon tape. Once again, pressurized gas was used to push the propellants from the tanks into the combustion chamber. Commercially available nitrogen gas cylinders were used for this purpose. But the regulators which came with the cylinders did not provide adequate flow of gas and had to be removed altogether. At the time, the ISRO engineers did not have pressure regulators which would bring the high pressure in the cylinder down to the operating level they needed. So they vented the gas till the pressure in the cylinder dropped suitably and then connected it up to the system.

The test stand, if it can be called that, consisted of a large metal plate standing vertically. The engine was mounted hortizontally with the nozzle facing away from the plate. When the engine was firing, the thrust from it would be taken by the plate. The propellant tanks were mounted behind the plate so that they would be protected if the engine exploded. Such explosions did, in fact, take place.

The tests were carried out in the yard of the Rocket Propellant Plant at Thumba. Initially, there were no electrically operated valves and the test began with someone opening the tap of the gas cylinder and then sprinting to safety. Others, taking shelter behind coconut trees, watched the firing of the engine. After electrically operated solenoid valves were developed in-house, the gas flow could be started safely from a distance by touching the electrical leads from the valve to a car battery.

Although these first attempts now appear bumbling, they served their pupose of providing an understanding of liquid propellants and the basics of engine design. They laid the foundations for the next two engines, both of which were pressure-fed engines.

The first of these was a 600 kg thrust engine. By this time, proper facilities for testing the engines on the ground were ready. Solenoid valves which could be opened and closed with an electrical pulse had been developed. Pressure regulators were available and gas could be stored at high pressure in special purpose gas bottles. Injector design, a key factor in liquid engines, had been improved. As a result, ISRO was able to build its first flightworthy stage based on liquid propulsion, not just an engine to be tested on the ground.

The *Annual Report* records, 'The launching of a 600 kg thrust liquid fuel rocket from Sriharikota in May 1973 was a landmark in the development of liquid propulsion technology.' The launch was unique in other ways too. While the liquid engine powered the second stage, the first stage used not one solid propellant block but several of them. Something like seven cordite propellant blocks of the sort used in the RH-75s were stacked inside a motor casing, with metal plates at either end to hold them in place. These blocks, burning furiously from both inside and outside, generated tremendous thrust, although for just a few seconds.

ISRO's liquid propellant engineers then decided to develop a 3 tonne thrust engine. The aim of this engine was more ambitious. The *1974-75 Annual Report* states that 'this engine, presently being developed for a test vehicle, will eventually be upgraded to meet the requirements of SLV-3 as a strap-on booster and as an upper stage'.

While the 600 kg engine had thirty injector elements, the 3 tonne had five times as many. Both engines had ablative linings to protect the walls of their combustion chamber. The asbestos-phenolic combination, tried out in the earlier 200 kg engine, was retained for the 600 kg engine and also in early trials of the 3 tonne engine. Later, an improved ablative lining, consisting of silica fibres embedded in phenolic resin, was substituted. A graphite piece provided protection at the throat, the constricted portion between the combustion chamber and the nozzle. The erosion of the graphite because of the hot gases was such that both engines could be operated for only about 45 seconds.

With the RH-560 as the first stage and the 3 tonne engine powering the second stage, this two-stage rocket was supposed to take a 175 kg payload to a height of over 100 km. The first flight, carried out in October 1976, was a failure. After the solid propellant in the first stage was exhausted, there was a brief period of no-thrust while the spent stage was separated and the upper stage ignited. But this no-thrust phase led to the liquid propellants floating away from the propellant outlets in the tanks. The liquid engine was unable to fire. For the next launch, the two stages were held in place by what is known as a vented interstage made up of metal struts, instead of a completely closed interstage. Both stages were ignited on the ground itself and the no-thrust coast phase was eliminated.

Subsequently, during 1978-79, the 3 tonne engine was flight-tested as a single-stage rocket. But sloshing and vortex development, disturbances which impede the smooth flow of liquid propellants, caused the engine to shut down prematurely.

The 3 tonne engine had also been designed with important improvements in mind. One was to use a gas generator to provide the pressurization needed to push the propellants from the tanks to the combustion chamber. The propellant tanks would be sufficiently pressurized at launch to take a little of the propellants to the gas generator. The propellants would burn in the gas generator and the hot gases produced would be fed back to the propellant tanks. The resulting pressure of hot gases would drive the propellants into the combustion chambers as well as continue to feed the gas generator. The *1977-78 Annual Report* says that the gas generator was successfully tested for a cumulative burning time of 150 seconds.

In addition, another gas generator had been designed which could be coupled to a turbopump. Pressure-fed systems are simple and hence highly reliable, but are suitable only for engines of lower thrust. For higher thrust engines, a turbopump is needed. In such systems, the hot gases from a gas generator drive a turbine, which, in turn, drives two pumps, one for each of the propellants. Turbopumps are complex machinery. As the parts rotate at high speed, they are designed, machined and assembled to high precision.

The speed of rotation can set up vibrations which tear the turbine apart. The turbine and pumps are usually mounted on the same shaft, and good seals are needed to prevent the propellants from leaking out and causing an explosion.

The turbopump for the 3 tonne engine was tested on the ground. ISRO involved external agencies, such as the National Aeronautical Laboratory in Bangalore, in carrying out these tests. Several annual reports refer to tests on the turbopump being carried out between 1977 and 1980.

But, neither a 3 tonne engine with gas generator pressurization, nor one with gas generator and turbopump were ever tested on the ground, let alone flown. By the end of the Seventies, technology for the French Viking engine had been acquired and indigenous efforts to develop an independent capability were shut down. The Viking contract will be discussed later in this chapter.

Liquid propulsion for the SLV-3

Since Scout had been chosen as the model for India's first launch vehicle, the role of liquids was only to maintain the launcher's orientation, called 'attitude' in aerospace jargon, along the three axes, pitch, yaw and roll. The aim of these attitude control mechanisms is to provide a force to either counteract a disturbance experienced by the vehicle or steer the vehicle along a desired trajectory.

In the Scout, two separate attitude control systems were used during the operation of the first stage. Jet vanes were positioned in the nozzle of the rocket motor. Swivelling the vanes would deflect the exhaust gases coming out of the motor, creating the correcting force needed. Jet vanes was one of the earliest attitude control systems employed in rocketry. The Germans used them in their V-2 missiles during World War II. The Scuds, such as those fired by Iraq during the Gulf War, also have jet vanes.

The Scout had four fins mounted at the bottom of the first stage. These fins provided aerodynamic stability. In addition, the tips of these fins could, like the jet vanes, be rotated. Such rotation would deflect the air stream and create the control force required.

The SLV-3 retained the Scout's fins and fin tip control. But although some attempts were made in developing vanes, the effort does not seem to have been seriously pushed. Difficulties in developing materials capable of withstanding the high temperatures is said to have been one major reason why jet vanes were given up. Instead, ISRO opted to develop a different technique called Secondary Injection Thrust Vector Control (SITVC). In a typical SITVC system, a fluid is injected at the nozzle into the flow of hot gases. The fluid sets up shock waves which deflect the flow. By changing the direction from which the fluid is injected as well as its quantity, the direction and magnitude of the control force generated can be adjusted. Thus a SITVC system has multiple inlets around the nozzle, with valves to control the amount of fluid injected.

The SITVC system was developed by the liquid propulsion group. Strontium perchlorate was chosen as the liquid to be injected. While fins and fin tip control were given up after the SLV-3, the first stage solid motor of every ISRO launcher as well as many of their strap-ons have been equipped with SITVC to control the vehicle's attitude.

The SLV-3's onboard guidance system had a complicated logic for using these control mechanisms. In the initial stages of flight, SITVC would be used till adequate speed built up for the fins and fin tip control to become effective. Thereafter, during the operation of the first stage, the fin tips would be used to maintain the vehicle's attitude. Only if these proved inadequate would the SITVC be resorted to.

The Scout rockets used hydrogen peroxide thrusters for attitude control during the operation of the second and third stages. However, the SLV-3 designers chose a slightly different path. The second stage attitude control system used bipropellant thrusters. Four thrusters, delivering 250 kg of thrust, provided pitch and yaw control while the motor was firing and another four thrusters, with 50 kg thrust, handled the same task during the coast phase when the second stage was being separated from the rest of the rocket. These 50 kg thrusters were later found to be unnecessary and removed in the last flight of the SLV-3. Another four small thrusters, giving just 10 kg thrust, were used for roll control.

The original plan had been to use aniline as fuel and RFNA as oxidizer, an obvious choice given the experience in using these propellants. But the RFNA-aniline combination was found to have limitations for use in control thrusters. In the meantime, hydrazine had been produced for the third stage thrusters and the RFNA-hydrazine propellant combination gave satisfactory results. A common set of tanks fed RFNA and hydrazine to all the thrusters. Pressurized gas, stored in separate gas bottles, fed the propellants when needed to the thrusters.

The third stage too used thrusters for attitude control, but this time they were monopropellant ones. Monopropellant thrusters use a catalyst to decompose the propellant, producing gas and much heat. For the SLV-3, parallel attempts were made to develop monopropellant thrusters using hydrogen peroxide (the propellant used in Scout) as well as hydrazine. The work on the hydrazine thrusters, using an indigenous iron-cobalt-nickel catalyst, progressed faster and these thrusters were ultimately chosen to fly on the SLV-3.

A liquid motor for APPLE

An interesting effort during this period was the attempt to develop a liquid propellant engine for the APPLE experimental communications satellite. The European Space Agency (ESA) carried this 600 kg satellite into space free of cost on the third qualification flight of the Ariane launch vehicle in June 1981; this was the first time that an Ariane launcher carried two satellites. As was explained in the previous chapter, the Ariane leaves the satellites in an elliptical geostationary transfer orbit. The satellite has to use its own propulsion capability (often referred to as the Apogee Boost Motor or ABM) to move to geostationary orbit.

At that time, most satellite ABMs were single or dual stage solids. Solids were seen as robust and reliable. It was the German company, MBB, which developed the first liquid propellant ABM for satellites. Liquids, as pointed out earlier, are more energy efficient and therefore less of the fuel needs to be carried to provide the same amount of thrust. The liquid engine's ability to stop and restart was also an important asset. Orbit correction could be carried out in stages, giving time to determine the satellite's orbit accurately and decide how much additional velocity it should be given. The result was that propellant wastage was minimized. A satellite's life is often determined by the amount of propellant it has for periodic corrections which are needed, using the onboard thrusters. Reducing propellant wastage in taking the satellite to geostationary orbit increases its life and thereby the financial returns from it. Today, all satellites in geostationary orbit have liquid ABMs.

According to R.M.Vasagam, project director for APPLE, when ESA first offered to accommodate a small Indian satellite on Ariane, they specified that the satellite could not weigh more than 150 kg. In addition, Ariane would leave the satellite in a considerably inferior transfer orbit. The apogee — the farthest distance between satellite and Earth — would be 17,000 km. The transfer orbit would be inclined at 17 degrees to the equator. By comparison, Ariane currently puts satellites in a transfer orbit with an apogee of 36,000 km and inclination of 4 degrees. The lower apogee and higher inclination meant that more fuel would be needed to take the satellite to geostationary orbit. With an already small satellite, a solid ABM was out of the question. Only the more energy-efficient liquids could do the job.

A liquid-propellant engine, generating 50 kg of thrust, was therefore designed and tested. It used RFNA and hydrazine as propellants. It was the first ISRO liquid engine to have regenerative cooling wherein one of the propellants is passed around the combustion chamber to cool it down before being injected into the same chamber. Instead of channels milled around the engine through which the propellant would pass, there were only metal buttons to maintain the space between the inner and outer walls. The engine is said to have been successfully tested for 1,000 seconds with a single hardware.

But this liquid engine was only a technology demonstrator. The development of a flight engine was not taken up because ESA had by then made it clear that Ariane would be able to put the satellite in a transfer orbit with an apogee of 33,000 km and an inclination of 11 degrees. They also agreed that the APPLE could weigh 600 kg at launch. Now the SLV-3's solid fourth stage could serve as APPLE's ABM. The solid motor was preferred to the liquid engine because development of the former was more advanced and, because of the SLV-3 project, its delivery within the specified time seemed more certain. The APPLE satellite had to be got ready within thirty months, points out Vasagam. The Europeans were so sceptical about India having a modern satellite ready within this period that ESA insisted that a structural model of the satellite be kept in France. If the actual satellite was not ready in time, the structural model would be carried instead. It would have been a considerable humiliation for ISRO if the structural model had been launched in place of APPLE, observes Vasagam.

If, however, the liquid-engine development had been pushed and was successful, India would have become one of the early users of liquid ABMs for its satellites.

Acquisition of Viking technology

ISRO's acquisition of technology for the Viking liquid engine used in the Europe's Ariane launch vehicle is one of the most extraordinary tales. ISRO didn't pay any money to buy the technology. Instead, ISRO placed several of its engineers to work alongside the French in the development of the engine. It also made and supplied a large number of pressure transducers, devices which permit pressure to be monitored through electrical signals.

The French space agency, Centre National d'Etudes Spatiales (better known as CNES), had been involved in Thumba from the very beginning. CNES technicians helped in the first launch of a sounding rocket from Thumba in November 1963. The following year, CNES signed an agreement with India for cooperation in space research. France supplied the Centaure sounding rockets for launch and also a radar to track sounding rockets in flight. Later India purchased the technology for making the Centaure sounding rockets from Sud Aviation of France.

Under Dhawan, ISRO's ties with CNES deepened. An ISRO-CNES Joint Commission was formed. At Dhawan's invitation, a seven-member delegation, led by the president of CNES, J.F. Denisse, came to India in January 1973 and visited ISRO establishments at Ahmedabad and Trivandrum. The first meeting of the joint commission was held on January 22. The joint commission set up two working groups, one on communications and TV satellites and the other on launchers.

The formation of the joint commission and establishment of a working group on launchers would have created an atmosphere conducive for collaboration between the Indians and the French. Through CNES, Muthunayagam and his team were able to get technical assistance from the French company Société Européene de Propulsion (SEP) for setting up the Static Test and Evaluation Complex (STEX) at Sriharikota. A convergence of interests deepened these bonds and led to a contract whereby Indian engineers would work on the development of SEP's Viking engine.

In 1973, an ISRO study group, headed by R.M.Vasagam, had submitted its report on the sort of launch vehicle ISRO might have to build after the SLV-3. This report will be looked at in more detail in a later chapter. Suffice to say here that the configuration recommended in the report would have involved a cluster of four liquid engines, each producing 60 tonnes of thrust, for the first stage and with one of those engines being used in the second stage as well.

The Europeans and the French had a similar requirement for their Ariane launcher. The European effort to have an independent launch capability goes back to the early Sixties. Ten European countries had come together in 1962 to establish the European Launcher Development Organization (ELDO) to develop the Europa launch vehicle. The Europa launcher had a British first stage, a French second stage, a German third stage and Belgian electronics. Three attempts to launch Europa-I from the Woomera range in Australia, between 1968 and 1970, all ended in failure. An improved Europa-II lifted off the CNES' launch facility in French Guiana in November 1971, only to wind up in the ocean a few minutes later.

Then Europa-III was proposed. Coming at a time when the Americans had landed men on the moon and were planning the space shuttle, doubts were increasingly expressed over Europa-III. Matters came to a head and to salvage the European alternative, the French proposed substituting the Europa-III with a less ambitious configuration based on proven technology. It received a lukewarm response and stiff negotiations were needed before the go-ahead was finally given by the European space ministers at a conference in the early hours of 31 July 1973.[1] The launch vehicle came to be called Ariane.

Based on experience gained by its predecessor, the Laboratoire de Recherches Balistiques et Aérodynamiques, SEP had developed the M40 engine, so named because it produced 40 tonnes of thrust, for the Europa-III. The M40 was the first turbopump-fed liquid engine developed by the French. As Europa-III evolved, SEP developed a more powerful version producing 55 tonnes of thrust and called it the Viking-1 engine. By the time the Ariane programme was approved, SEP was already working on the Viking-2 which would generate 60 tonnes of thrust. The Viking-2 was tested for the first time in December 1973.

As in the configuration suggested by the Vasagam Committee, the Ariane-1 had four 60 tonne thrust liquid engines in its first stage and a similar engine in the second stage. Two Viking engines were derived from the Viking-2 for the Ariane-1 programme. The Viking-4 would be the single engine for the Ariane second stage. A cluster of four Viking-5 engines would be required for the Ariane first stage. It was against this background, when the French were working to develop the Viking-4 and 5, that the agreement with ISRO was negotiated and signed.

After the first meeting of the ISRO-CNES Joint Commission and the formation of the working groups in January 1973, there was a flurry of activity. G. Boelle, CNES' deputy director-general for International Affairs, came to Bangalore in October 1973. A CNES team visited various Indian industries in December that year and Dhawan went to France in February 1974. The second ISRO-CNES Joint Commission meeting was held in Paris in July 1974. Shortly afterwards, ISRO negotiated and signed the agreement with SEP.

India would send its engineers to France where they would work alongside SEP specialists and provide hundred man-years of effort towards developing the Viking engines needed for Ariane. In addition, ISRO agreed to make and supply several thousand pressure transducers to SEP. In return, the engineers would be able to acquire the capability for making the Viking engine in India.

The contract was a product of India's inability to pay hard currency for know-how, says the man who signed the contract on behalf of the Department of Space. This was none other than T.N. Seshan, then joint secretary in the department, who would later become India's chief election commissioner. But why did the French agree to what was practically a barter arrangement? 'What motivated them is something I can't answer,' says Seshan. There was tremendous goodwill among the French towards the Indian space programme, he adds.

But there may well have been other more compelling reasons for the French to agree to such an arrangement. After the problems with Europa, the other European countries were far from enthusiastic about spending more money on developing a launcher. Even with France putting up three-quarters of the money required, the European ministers had only grudgingly agreed to the Ariane programme. Ariane had to fly within six years of the proposal being approved. The Viking engines had therefore to be developed within tight budgetary and time constraints. Delays or cost overruns could scuttle Ariane. With so much at stake, the French were probably more favourably inclined towards an agreement which provided trained manpower from India to work on the Viking's development.

For ISRO, it was a windfall. In just five years, it got the technology and could therefore take a big leap forward to a modern high-thrust, turbopump-based liquid propellant engine. Nor was it a conventional technology transfer consisting of a bunch of drawings along with instructions for manufacture, assembly and testing. ISRO sent some fifty of its engineers to France. Working alongside their French counterparts, these engineers picked up the design principles of the Viking engine and stage. As one of ISRO's senior liquid propulsion engineers remarked, 'we didn't get just the know-how, but also the know-why'.

Of the hundred man-years of effort which the Indian engineers would provide under the contract, SEP could utilize seventy-five man-years as they wished. But the contract is said to have given ISRO the freedom to use the remaining twenty-five man-years in areas where it wanted to acquire capability. The contract reportedly provided for technology transfer to India in the form of drawings and other technical documentation for the engine alone. But having worked in France, the Indian engineers learnt whatever else they needed to design and make a stage using the engine. By the time they came back to India, they had had hands-on experience in all aspects of liquid propulsion, from design issues to assembly, fabrication and testing. These skills and the understanding they had gained were vital in building the engine and stage in India.

I was told that by the time the ISRO engineers returned in 1980-81, they had given SEP something like 135 man-years of work. By mid-'76, ISRO had established the Pressure Transducer Unit at Bangalore and the first batch of transducers were sent to France. ISRO planned to produce and supply over 7,000 transducers by the early Eighties. The Vikas project, as it was called, had a sanctioned cost of Rs 2.7 crore, covering both the expense of keeping engineers in France and supplying the transducers. By the end of 1980-81, the total expenditure which ISRO actually incurred was around Rs 4 crore. The contract with SEP had given ISRO not just a comprehensive technology transfer, but one made available at a ridiculously low price. It was the ultimate bargain.

Major challenges now lay ahead. The next generation of ISRO launch vehicles would use liquid propulsion for the core stages, not just for attitude correction. The Indian equivalent of the Viking engine had to be successfully fabricated in India and a suitable stage designed and built around this engine. The experience gained with pressure-fed engines too would be required, both for launch vehicles as well as for the Insat satellites.

CHAPTER

6

The ASLV: A Technological Bridge

THE SLV-3 WAS a very basic launch vehicle, a vital first step in developing the technological and project management capabilities required to attempt more powerful launch vehicles. These, in turn, were needed to fulfil Sarabhai's vision of not only having the capability to build operational satellites but also for launching them.

A committee was set up towards the end of 1972, soon after Dhawan took over as chairman, and studies began on the sort of launch vehicle ISRO needed in order to launch operational satellites. These studies continued all through the Seventies while the development of the SLV-3 was going on. By the mid-Seventies, it was clear that ISRO would be building two types of operational satellites: the Indian Remote Sensing (IRS) satellites for remote sensing applications and the Insat satellites for communications, direct TV broadcasts and meteorology.

The evolution of the launch vehicle configuration during this period will be discussed in a later chapter. Suffice to say here that, by the time of the SLV-3 launch, it had been decided that the next step would be to develop the Polar Satellite Launch Vehicle (PSLV) to put the IRS satellites into orbit. The PSLV could then become the basis for developing the more powerful launchers needed to carry the Insat satellites.

The issue then arose as to whether to embark straightaway on the PSLV or have an intermediate launch vehicle between the SLV-3 and the PSLV. It was definitely a big technological jump from the SLV-3 to the PSLV. An intermediate vehicle would allow some critical technologies needed for the PSLV to be tested more cheaply. It would also give ISRO continued visibility during the ten years or so needed for the PSLV development. Against these arguments were worries about whether ISRO could pursue three launch vehicle programmes at the same time: the SLV-3 continuation programme, development of an intermediate launch vehicle and development of the PSLV. Ultimately, ISRO took the view that the benefits of an intermediate launch vehicle outweighed its disadvantages.

The result was the Augmented Satellite Launch Vehicle (ASLV). The ASLV was particularly appealing because it appeared to be no more than a straightforward augmentation of the basic SLV-3. The *1980-81 Annual Report* stated:

> The main goal of this project is to achieve in about two to three years time an augmented satellite launch vehicle based on the SLV-3 as the core with minimum modifications, but capable of placing a 150 kg payload in near earth orbit from SHAR.

The four stages of the SLV-3 would form the core of the vehicle, with two strap-on motors, each identical to the SLV-3's first stage, attached to either side of the ASLV's first stage. The ASLV would have a more sophisticated onboard guidance system as well as a bulbous heat shield so that satellites larger than the vehicle's diameter could be accommodated. As it was capable of carrying 150 kg satellites, ASLV's payload capability was more than three times greater than that of the SLV-3. Its better guidance system would allow the ASLV to achieve the intended 400 km near-circular orbit with greater precision.

Not only would the ASLV be a technological stepping stone to the PSLV, but could also, ISRO believed at one time, be used as a low-cost launcher to put small scientific and experimental satellites into orbit. 'Once proven, ASLV will serve as a workhorse vehicle for all Low Earth Orbit missions of ISRO,' said an ISRO brochure on the ASLV published in November 1986.

Since it used the same core stages as the SLV-3, ISRO assumed that development of the ASLV would be quick and uncomplicated. The ASLV and PSLV projects were both cleared in June 1982. The ASLV's sanctioned project cost at the time was Rs 19.73 crore and the first developmental flight was scheduled for 1985.

ISRO speedily learnt its error. The ASLV was not a simple variant of the SLV-3, but a totally new beast, and a particularly recalcitrant one at that. Two successive failures just a year apart plunged the launch vehicle teams into the depths of despair. One more failure could have jeopardized the still fledgling launch vehicle programme.

Instead of the three years it had so confidently forecast, ISRO spent nearly a decade before successfully launching the ASLV. Unlike the failure of the first SLV-3 which was caused by a minor problem, there were no quick-fix solutions for the ASLV. The ASLV suffered from some basic design problems arising out of inexperience. The Indian launch vehicle teams emerged from this chastening 'agni pariksha', or trial by fine, with a deeper understanding of the complexities of launch vehicle design.

A relatively easy launcher?

The initial development of the ASLV right up to the first launch was strewn with the usual complement of difficulties and problems. But it did not throw up any unpleasant surprises which could bottleneck and delay the whole development programme.

One of the major design challenges for the ASLV team had to do with the addition of the strap-on motors. This was one of the key technologies needed for the PSLV also. The ASLV would have two strap-on motors attached to either side of the first stage. The problem with adding strap-on motors was that if one of the strap-ons produced too much or too little thrust, it would push the launch vehicle to one side or perhaps even send it spinning out of control. In order to achieve similar performance, ISRO engineers developed techniques of simultaneously casting segments for both strap-ons.

There were also questions related to the stresses and strains which would be created by air flowing over the strap-ons as the launch vehicle quickly accelerated through the atmosphere to speeds much above that of sound. A major worry was whether on separation, after their solid fuel was exhausted, the strap-ons would fall clear of the launch vehicle or collide with it. These questions had to be

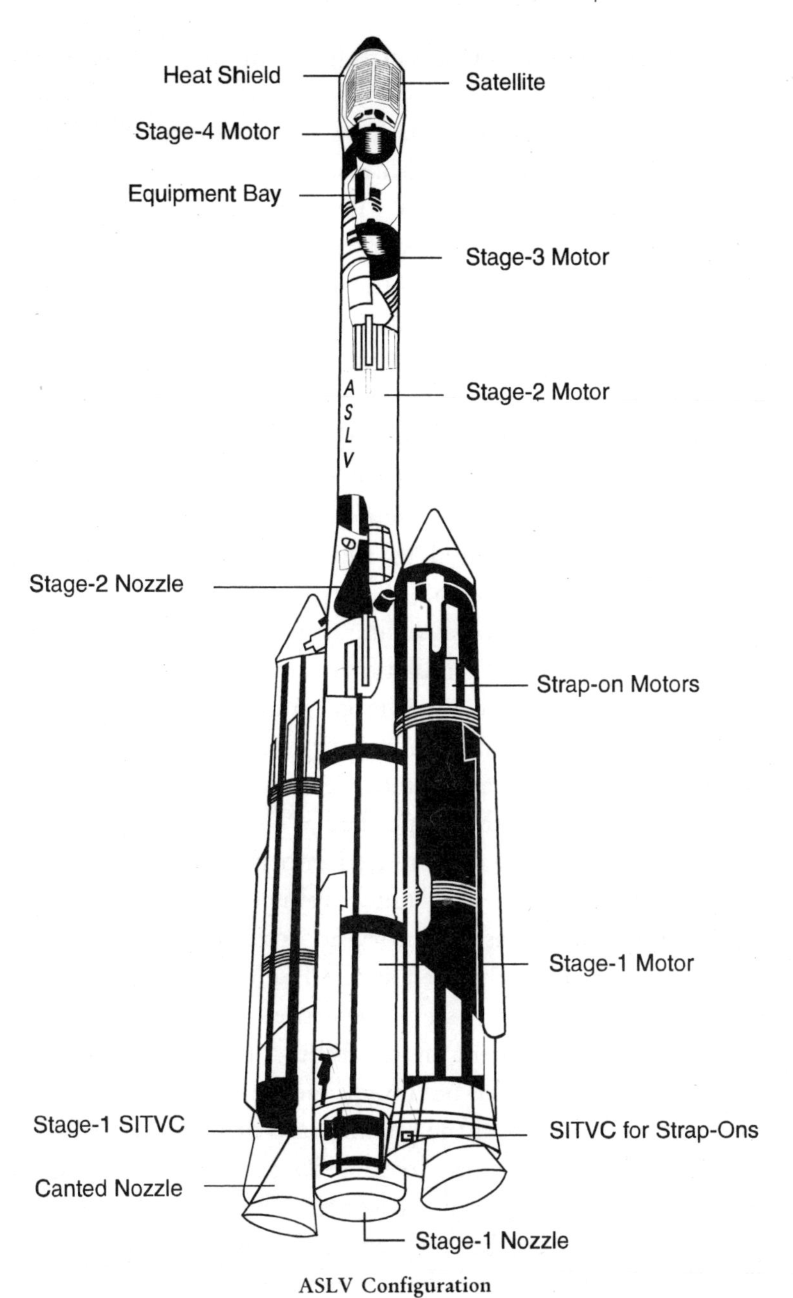

ASLV Configuration

resolved through wind tunnel tests, computer simulations and ground tests.

Before the strap-on technology was flown in the ASLV, it was first demonstrated with a sounding rocket. An RH-300 sounding rocket, with two RH-200s strapped on either side of it, was tested from Sriharikota in October 1985.

The bulbous heat shield was another important technology needed for the PSLV and attempted for the first time in the ASLV. In the SLV-3, the heat shield was no wider than the fourth stage. A bulbous heat shield could accommodate satellites bigger than the upper stage. Its disadvantage, however, was that it would disrupt the smooth airflow over the launch vehicle. The turbulence it created as the vehicle accelerated to supersonic speeds could be dangerous. Since the shape of the heat shield was critical, information from published literature as well as extensive wind tunnel tests and simulations went into its design.

The SLV-3 had an 'open loop' guidance system which did no more than carry out a preordained sequence of operations and keep the launch vehicle correctly oriented. The ASLV, on the other hand, had a 'closed loop' guidance system. The guidance system used information from special sensors to determine the launch vehicle's position during flight and adjusted its trajectory accordingly.

Though not as advanced as the PSLV guidance system, the ASLV guidance system nevertheless represented an important improvement in capability. The PSLV's closed loop guidance would be much more sophisticated and control the injection of the satellite into orbit. In the ASLV, on the other hand, the closed loop guidance would end when the third stage was jettisoned. The early part of the ASLV flight would be with an 'open loop', enabling the launcher to follow a predetermined trajectory. The closed loop guidance began only with the operation of the second stage. It would then attempt to modify the trajectory, depending on the performance of the second and third stages, so that the satellite could be put as close to a 400 km circular orbit as possible.

The SLV-3 had been integrated horizontally in a separate building. The fully integrated launch vehicle was taken on a special trailer to the launch pad and then raised to the vertical position, unlike the PSLV which would be integrated vertically right on the launch pad to minimize stresses on the launch vehicle. A Mobile Service Tower (MST) would cover the launch pad during integration, providing protection from the weather as well as the facilities required for stacking the stages. The MST would be moved clear of the launch pad just a couple of hours before launch. This concept of integrating the launch vehicle vertically on the pad was tested out in the ASLV.

The first launch of the ASLV

In March 1987, the ASLV at last stood on the launch pad at Sriharikota, ready for its maiden flight. There had been no hitches during the assembly of the stages. The launch campaign and the final countdown which began thirty-one hours before launch went exceptionally smoothly. At 12.09 p.m. on March 24, the ASLV's two strap-on motors were ignited and the launch vehicle soon lifted clear of the launch tower. In the first few moments, the launch vehicle appeared to follow the planned trajectory perfectly.

The first sign of trouble came 50 seconds after the launch. The altitude achieved by the ASLV was only 9.82 km instead of 10.09 km, recalls S.C. Gupta, who had by then taken over from Gowariker as director of VSSC. 'The next major event was the ignition of the first stage,' he susequently said in an interview given to *Countdown*, the VSSC house journal. 'On the monitor one could see two exhaust trails from the strap-ons and a third one which ought to be from the first stage but which appeared too feeble to be so. Nevertheless I expected this feeble trail to intensify soon but, alas, it didn't.'

The first stage had failed to ignite and the vehicle tumbled earthwards, out of control. The launch vehicle broke up in mid-air and the various pieces ultimately splashed down in the Bay of Bengal.

An internal Failure Analysis Committee was set up and asked to submit its report in less than a month. The launch vehicle had continuously radioed down a large number of operating parameters as it flew. It was also accurately tracked by radar from the ground. In addition to this data, a number of simulation studies were also carried out to try and understand the cause of the failure. It was found that most parameters, such as the vibrations experienced by the launch vehicle, were within permissible limits.

The Failure Analysis Committee, headed by R. Aravamudan, then associate director of VSSC, stated that 'the chief cause of failure of the mission was non-ignition of the first stage'. But the committee was unable to satisfactorily explain the reason for this, although it had examined in detail thirty-seven ways in which such a failure could have happened.

The committee took the view that non-ignition of the first-stage motor could be explained by an inadvertent short-circuit in both the primary and back-up ignition circuits, both ignition circuits being inadvertently open, or a random malfunction of the safe/arm device. Several people in ISRO point out that a short circuit, open circuit or random malfunction would hold true for the failure of any electrical circuit.

The safe/arm is a mechanical device which in the 'safe' position prevents the igniter from accidentally firing the solid motor. The safe/arm has to be in the 'armed' position for the igniter to work properly. There was no evidence to show that the safe/arm had malfunctioned during the ASLV flight. Indeed, when safe/arms made in the same batch as the one which flew on the ASLV were rigorously tested for the failure analysis, they functioned correctly.

The then ISRO chairman, U.R. Rao, is said to have been convinced that the failure was due to a problem with the safe/arm. The committee's finding of a 'random failure' of the safe/arm, which can neither be proved or disproved, is believed to have been an attempt to accommodate the chairman's views.

Although the Failure Analysis Committee had not identified any specific cause for the first stage not igniting, it did suggest various changes. These included removing the mechanical safe/arm and replacing it with an electrical safe/arm of the sort used in SLV-3. A mechanical safe/arm is, however, a more foolproof method of preventing accidental ignition of a solid motor. The PSLV uses mechanical safe/arms, although of a very different design.

The ASLV's second flight

The recommendations of the Failure Analysis Committee were faithfully implemented. 'Only that way can we ensure success of the next flight of the ASLV,' said an article in *Countdown*.

On 13 July 1988, the second ASLV's twin strap-ons were ignited and the launch vehicle lifted off from

Sriharikota at 2:48 p.m. Both strap-ons performed normally till their fuel was exhausted. Unlike in the previous flight, the first stage then ignited. But the launch vehicle had already begun to tilt and it was soon beyond control. The ASLV broke up and ended, as before, in the Bay of Bengal.

It now became abundantly clear that no quick and easy fixes were possible. Very thorough investigations and reviews were needed if the ASLV programme was to be rescued. Apart from an internal Failure Analysis Committee headed by S.C. Gupta, a national Expert Review Panel, headed by R. Narasimha, was asked to examine the failure. This time, the failure analysis was thorough and able to put its finger accurately on the ASLV's design limitations.

The fundamental issue in the ASLV was of not allowing the vehicle to tilt and go out of control during the critical transition from the strap-on motors to the first stage. The SLV-3 had four large fins at the base of the first stage which increased its stability. These fins had been removed in the ASLV in the belief that adequate control was available through other means, such as Secondary Injection Thrust Vector Control (SITVC is explained in the previous chapter). But the SITVC works by altering the direction of thrust from the solid motor to correct the vehicle's orientation. Therefore, unlike the fins, the effectiveness of SITVC depended on the amount of thrust being generated by the solid motor.

In the ASLV, the transition from the strap-ons to the first stage occurred at a height of just 10 km. By comparison, the SLV-3 would be at a height of about 30 km when the transition from the first stage to the second stage occurred. In the ASLV, the burning out of the strap-ons, their separation and then the ignition of the first stage occurred at a time when the dynamic pressure was very high and winds and gusts extreme. In such an environment, it becomes particularly important to minimize periods when control forces may not be adequate enough to hold the vehicle's orientation. If a launch vehicle tilts too much, the loads on it become very high, increasing the risk of vehicle break-up.

In the second flight, the strap-ons burnt out about 1.5 seconds earlier than expected. Consequently, the thrust from these motors began tailing off sooner, thereby reducing also the effectiveness of the SITVC. Unfortunately, the winds and gusts were also more severe during the second ASLV flight than in the first. As the control capability of the SITVC fell off, the strong winds and high dynamic pressure began turning the vehicle violently sideways. For a period of three to four seconds, there was not enough control force available to correct this unstable vehicle's orientation. By the time the first stage fired and its thrust (and the SITVC's effectiveness) built up, it was too late. The launch vehicle had gone beyond control and soon broke up.

The Expert Review Panel pointed out that as a result of its limited flight experience, ISRO had not fully recognized the impact that higher winds and early motor tail-off would have on the launch vehicle. These inadequacies had not been revealed in the first ASLV flight because its failure had been governed by the non-ignition of the first stage. Moreover, the strap-ons had functioned for a longer period and the winds conditions were relatively benign on that occasion.

The Expert Review Panel recommended that 'recognising the inherent dispersion in the burn out of strap-on boosters, the ignition of the core [first stage], instead of being at a prefixed time, should be preferably linked to the event when the strap-on boosters become ineffective in the tail off region'. It also suggested that improvements be made to the control systems to take care of worst-case disturbances and that the autopilot in the guidance system be suitably redesigned.

ISRO's launch vehicle teams took these lessons to heart. As the panel had suggested, periods when no control forces were available to keep the launch vehicle on course, especially in the turbulent lower atmosphere, were minimized. Pressure transducers were installed inside the strap-ons so that they could detect the tapering off of thrust. If, as happened in the second flight, the strap-ons burnt out earlier than expected, the transition to the first stage would be correspondingly advanced. Similarly, the transition sequence from the first to second stage would be initiated when onboard sensors detected a fall in acceleration.

The autopilot software was redesigned to cope with high winds even when the strap-ons burnt out early. The control system too was improved. The vehicle structure was strengthened so that it could tolerate more strain. The composition of the solid fuel in the two strap-ons was modified to provide lower thrust, and hence lower dynamic pressure, during most of their operation.

Later, it was decided to restore two fins to the ASLV first stage. These fins are said to have improved the vehicle's stability.

ASLV failure improves PSLV design

These insights were also applied to the PSLV, then under development. The PSLV was, however, reassuringly different from the ASLV. Most of its thrust, right from launch, was provided by its huge solid first stage, with the strap-ons only augmenting this thrust. Since the first stage would still be firing when the strap-ons burnt out and separated, control forces would be available throughout the launch vehicle's flight through the lower atmosphere.

Although the maximum dynamic pressure experienced by the PSLV would be less than 70 per cent of that experienced by the ASLV, ISRO reduced it still further. Instead of the original plan of igniting four of PSLV's strap-ons at launch and the remaining two later on, the sequence was reversed and only two strap-ons would be ignited on the ground along with the first stage. This would reduce the maximum dynamic pressure by 30 per cent.

In the PSLV, the burn-out of the first stage and transition to the second stage occurs at an altitude of about 76 km where winds and dynamic pressure would be far lower. Although a longer period without control forces could be tolerated, ISRO took no chances. The process of firing the PSLV's second stage, powered by a liquid engine, would be initiated as soon as the fall in acceleration showed that the first stage was nearing the end of its operation.

In the light of the ASLV experience, PSLV's autopilot was improved and the control forces increased so that they could cope with much higher winds.

Success at last

ISRO became almost paranoid about the controllability of the ASLV. It is true that a third failure would have had disastrous consequences for the entire launch vehicle programme. Even so, sometimes it seemed that the studies and simulations would never end. Many joked that the third ASLV would be a flying control system. Many of the changes added to the launch vehicle's weight and reduced its payload by about 40 kg. By now, however, payload, which seemed so important when the launch vehicle was conceived, could be sacrificed in the interests of achieving a successful flight.

Finally, on 20 May 1992, the ASLV was launched for a third time from Sriharikota. This time, after an uneventful flight, it put the 106 kg SROSS-C1 satellite into orbit about 450 km above the earth. Instead of the wild jubilation which greeted the successful launch of SLV-3, there was heartfelt relief all round. ISRO had successfully crossed an important Rubicon.

In May 1994, just four months before the launch of the first PSLV, the ASLV was successfully launched once more. It put the 113 kg SROSS-C2 satellite into an elliptical orbit of 938 km by 437 km. With this launch, ISRO drew down the curtains on the eventful ASLV programme.

ASLV's contribution to the launch vehicle programme

For all the heartbreaks it caused, the ASLV was an important watershed in ISRO's understanding of launch vehicle design. ISRO became acutely aware of maintaining control of the launch vehicle during its flight through the lower atmosphere where there could be unexpectedly high winds. Control of the launch vehicle was particularly important during the transition from one stage to the next. Such transitions had to be even more carefully managed when solid stages were involved and could burn out earlier than expected.

It is fortunate that ISRO was able to learn these lessons with the ASLV rather than the more complex and expensive PSLV. Each ASLV cost less than Rs 10 crore (at 1989 prices when the third and fourth ASLV launches were sanctioned). The PSLV would cost many times more. The PSLV's relatively smooth road to success owes much to the ASLV experience. The lessons learnt from the ASLV and incorporated into the PSLV made the latter more robust, says G. Madhavan Nair, who was the project director for PSLV during its early flights.

Both ASLV failures occurred around the same time when the strap-ons burnt out and the first stage was to take over. Could the second ASLV failure have been avoided if the first had been more thoroughly investigated? Several senior ISRO launch vehicle specialists say that the analysis of the first ASLV failure was not as thorough as it ought to have been. But they also point out that the issues of controllability, which became so starkly evident in the second failure, were not obvious in the first failure. That failure quite simply appeared to be a case of the first stage motor not igniting. There was widespread feeling — maybe even hope — that the ASLV's failure was due only to some simple problem for which a quick fix was possible.

Success, it is said, has many fathers while failure has none. M.S.R. Dev, the ASLV project director, and his team were often severely criticized — some say even humiliated — for their handling of the ASLV development and its associated problems. But ASLV's eventual success and its unquestioned contribution to ISRO's understanding of launch vehicles did not, unfortunately, earn them the recognition they deserved for coping with one of the most trying phases in the development of launch vehicles in this country.

CHAPTER

7

Guiding a Launcher From Ground to Orbit

Unlike a sounding rocket, a launch vehicle requires an advanced guidance and control system in order to inject a satellite into orbit. A sounding rocket has only to carry its payload straight up till its fuel is exhausted and then it falls back to earth. It does not need a complicated system for stabilizing and guiding it. A sounding rocket's fins give it stability. This stability is increased by setting the rocket spinning soon after it leaves the launcher. A spinning body resists the axis of its rotation being moved. The combination of fins and spin stabilization suffices for a sounding rocket.

But the purpose of any launch vehicle is to put a satellite (or some other payload) into a particular orbit. The sort of orbit will vary depending on the application. Remote sensing satellites may need, for instance, to be in a polar

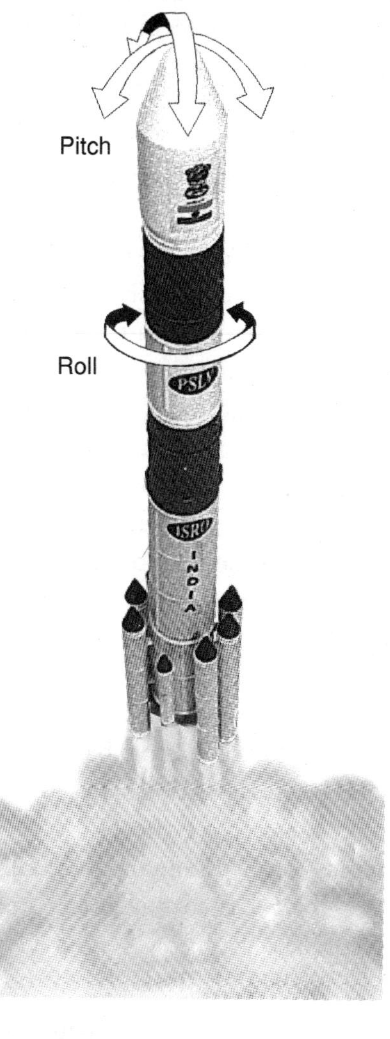

sun-synchronous orbit. In this type of orbit, the satellites come over a place at the same local time on each occasion so that lighting conditions remain constant.

Large communication satellites have to be placed in an elliptical geostationary transfer orbit. These satellites then use their onboard rocket motors to move into geostationary orbit. The greater the accuracy with which a launch vehicle can put a satellite into orbit, the less fuel the latter has to expend in correcting the orbit and the longer its operational life.

Simple spin stabilization will not do for a launch vehicle. The orbit a satellite achieves depends on the height, velocity and angle at which it is injected by the launch vehicle. Therefore, while a launch vehicle's solid and liquid stages provide the brute force needed, onboard systems are required to guide the launch vehicle along a suitable trajectory. To do this, the vehicle's orientation, or attitude, to use the technical term for it, has to be controlled at all times along three mutually perpendicular axes, pitch, yaw and roll.

The onboard guidance and control system has to constantly monitor and correct the launch vehicle's orientation. The guidance and control system has to be able to cope with winds in the turbulent lower atmosphere as well as any other disturbances which occur during the flight. Such events can push and rotate the launch vehicle. If the vehicle's attitude isn't quickly corrected, it can tumble out of control and break up. In the second ASLV flight, for instance, high winds coupled with lacunae in the control system, led to the failure of the mission. The first flight of the Polar Satellite Launch Vehicle (PSLV) failed after the disturbance created by two stages touching each other during separation could not be corrected.

At the same time, in order to take the launch vehicle along the curving trajectory needed to achieve the final orbit, the vehicle's attitude will also have to be deliberately changed. Usually, the change is made only along the pitch axis. The PSLV is probably unique in carrying out a major change in the yaw axis as well. As will be explained in the next chapter on the PSLV, in order to avoid spent stages falling on Sri Lanka, the PSLV is launched south-eastwards and then turned due south in mid-flight.

How inertial guidance and control is achieved

The onboard systems required for this purpose have three principal parts:

1. *Sensors*

   Two sorts of sensors are needed. These are the gyros (an abbreviation for gyroscopes) and the accelerometers which give information on the launch vehicle's attitude and acceleration respectively.

2. *Navigation, Guidance and Control (NGC) System*

   The data from the gyros and accelerometers is processed by the NGC to calculate the vehicle's current attitude and position. It then decides whether any correction or changes are needed. The launch vehicle's position is fixed and the course it should follow determined in terms of a fixed frame of reference, with the launch pad as its origin. This frame is termed the 'inertial frame' and this system of guiding the launch vehicle using only onboard equipment as 'inertial guidance'. The gyros and accelerometers are known as 'inertial sensors'.

3. *Control plants*

   The correction and changes in the launch vehicle's attitude are carried out using the various control plants. These systems generate forces to alter the vehicle's attitude so as to steer it along the desired trajectory or correct the attitude in the face of a disturbing force. Control plants used in the ISRO launch vehicles include:

   - Fin tip control: The SLV had four large fins fixed to the base of the first stage, the tips of which could be rotated. Rotating the fin tips diverts the flow of air over them. This generates the control forces needed.
   - Direction of thrust: Control can also be exercised by altering the direction of thrust from the main propulsion system. Such control plants are widely used.
     - All first stage solid motors in ISRO launch vehicles have had Secondary Injection Thrust

Vector Control (SITVC). A small quantity of fluid – ISRO uses strontium perchlorate – is injected at the nozzle of the solid motor into the stream of hot gases. This generates a shock wave which deflects the gas flow and thus provides forces for control.

- Alternatively, solid motors can be equipped with a flex nozzle, as in the PSLV's third stage, which can be swivelled.
- Liquid engines, including the Vikas engine used in the PSLV's second stage and the two liquid engines in its fourth stage, can be 'gimballed' and swivelled when needed.

- Thrusters: These are liquid engines, using either monopropellants or bipropellants. They can be switched on and off on command. Such thrusters were used for attitude control in the second and third stages both in the SLV-3 and later in the ASLV. Two liquid engines are used for roll control during the operation of the PSLV's first stage.

Development of control plants to correct and change a launch vehicle's attitude have been covered in an earlier chapter. Therefore only the development of inertial sensors and inertial guidance will be examined here.

The building blocks: gyros and accelerometers

Gyros give accurate information about the change in a body's orientation, be it a launch vehicle or a satellite. Mechanical gyros use the fact that the axis of a spinning body moves in a predictable fashion (called 'precession') when perturbed. By measuring the precession, the gyro shows how fast the launch vehicle is turning along one or more axis. Since it is important that the rotor's axis precess freely, minimizing friction, which would inhibit such movement, is important in mechanical gyros.

The accelerometer measures acceleration by taking advantage of the inertia of a body, much as passengers would be pushed back into their seats when the vehicle they are travelling in suddenly accelerates. I have not attempted to go into the design intricacies of these inertial sensors. The book *Inventing Accuracy: A Historical Sociology of Nuclear Missile Guidance* by Donald MacKenzie, published by the MIT Press, gives a highly readable account of how gyro and accelerometer design evolved.

The technology for gyros and accelerometers is among a country's most closely-guarded secrets. These inertial sensors are used in the guidance of both launch vehicles and missiles. Better inertial sensors improve the accuracy with which a launch vehicle can put a satellite into orbit. When used in missiles, these sensors increase the accuracy of the warheads. As a consequence, even the performance of such sensors is usually classified information.

S.C. Gupta, who had overall charge of the development of both inertial sensors and the guidance system, pointed out in a talk in 1980:

> The governing theory of operational gyroscopes and accelerometers [is] well known and well documented... However, the technological information about their construction features and the exotic materials used, which make them capable of measurement with the accuracies required during the flight is highly classified... As these devices form the critical subsystem of guided missiles, of both tactical and strategic calibre, they are classified under weapon systems of the most sophisticated type. As such,

these devices are not generally available for even the peaceful uses of outer space.

Developing the gyros and accelerometers needed by the Indian space programme, both for satellites and for launch vehicles, has been the task of a group which was now been removed from its parent organization, VSSC, and made into a separate unit, the ISRO Inertial Systems Unit (IISU).

Around 1969-70, a rate gyro was built and demonstrated to Sarabhai. 'At that time, we didn't even have a bench lathe inhouse,' remembers Gupta. All the precision machining and other fabrication had to be got done elsewhere. Even the turntable, to rotate the gyros to see if they responded correctly, had to be put together from bits and pieces got from the Air Force. The idea was, however, only to demonstrate that gyros and other inertial systems could be designed and built in India. Sarabhai was impressed.

At that time, the Hindustan Aeronautics Limited (HAL) had tied up licence agreements with various European companies and is said to have offered to set up licence production of the inertial sensors needed for the space programme if they would provide Rs 1 crore. But Sarabhai preferred the indigenous development route and the Atomic Energy Commission gave Rs 1.25 crore to ISRO to establish what came to be the Precision Instruments Laboratory (PIL). Gupta and N. Vedachalam, the IISU director till recently (and currently director of the Liquid Propulsion Systems Centre), visited Switzerland and other countries. They bought half a dozen high precision fabrication and inspection machines for the new facility.

In the initial stages, a variety of specialized components required for building the inertial sensors were made and tested. But it was quite clear that inertial sensors of the quality required for launch vehicles were not going to be ready in time for the SLV-3.

The open-loop guidance chosen for the SLV-3 demanded only information about the changes in the launch vehicle's attitude, not calculation of its position. So a system which used three gyros measure changes in the three axes sufficed. Sagem of France supplied the Inertial Measurement Unit (IMU) for the first two flights of the SLV-3. ISRO had approached US companies, but they refused to provide even catalogues about their gyros, says Gupta.

The IMU consisted of three gyros in a stabilized platform arrangement. The stabilized platform has gimbal mechanisms and tiny motors so that its own orientation along the three axes can be precisely altered. When the gyros detect a change in attitude, there is an electronics circuit which sends a signal to rotate the platform back to its original position. Thus, no matter how the vehicle tilted, the platform would be returned to its original orientation in respect to the inertial frame.

The stabilized platform system has two major advantages. The stabilized platform, by isolating the gyros from the launch vehicle's vibrations, provides a more benign environment and the gyros in this system need not be so rugged. The Scout, on which the SLV-3 was modelled, used a strapdown system where the gyros were fixed directly to the launch vehicle frame. According to Vedachalam, it was not possible to get strapdown quality gyros for the SLV-3. Since the stabilized platform maintains constant orientation with respect to the inertial frame, it also has the advantage of reducing the onboard computation load, an important factor when high speed space-quality processors were not available and analog electronic circuits had to be used.

The Sagem IMUs flew on the first two flights of the SLV-3. Subsequently, ISRO imported Sagem gyros, integrated them with a stabilized platform of its own making and supplied these IMUs for the next two flights.

The stabilized platform for the indigenous IMU was made larger than the imported one to accommodate the larger Indian-made gyros and for adding accelerometers at a later date.

The Indians wanted to build a gyro which could be used both in a stabilized platform as well as in the strapdown mode. To do this, they opted for the floated gyroscope design where the gyro floats in a fluid to reduce friction. This design, pioneered by Charles Draper and his Instrumentation Laboratory at MIT, senses the vehicle's rotation along only one axis. The idea was to first build a rate gyro and then use that experience to build a rate-integrating gyro. In the rate gyro, the precession of the spinning rotor's axis indicates the rate of turn. The rate-integrating gyro has a tiny motor to rotate the axis back after it has precessed. It is thus able to indicate the angle through which the vehicle has turned, rather than the rate of turn.

The Indian group faced formidable problems in the development of these gyros, as indeed have other countries when developing inertial sensors. One was the unavailability of beryllium, which was not being produced within the country and which could not be imported either. Beryllium's light weight gives it a strength to weight ratio more than twice that of steel. It has considerable dimensional stability and can retain its size and shape with considerable accuracy over time. It was only in 1977 that the departments of Atomic Energy, Space and Electronics agreed to jointly establish a beryllium production plant. The unit, which was built on the outskirts of Bombay, became operational in November 1982. Since beryllium is toxic, the plant incorporated suitable precautions to prevent escape or inhalation of its powder, gas or fumes. For the same safety reasons, a beryllium machining facility was established right next door and the parts for the gyros were machined there.

Till beryllium parts became available, it was decided to try and build the gyros using aluminium. However, since the density of aluminium is greater than that of beryllium, more liquid was needed so that the gyro would float. This made the complete gyro package much larger. In addition, the fluid used in floated gyros, fluorolube, was being made by only one company in the world and could not be procured. Substitutes were tried, but were not as good. Even assembling the gyros and filling the fluid was an intricate exercise.

The net result seems to have been that the Indian floated gyros lacked the performance needed for the inertial navigation of launch vehicles. The Sagem gyros were reportedly used for the ASLV as well. While efforts to improve the floated gyros continued, ISRO began work on an entirely new gyro, the dynamically tuned gyro or DTG. When the DTG development was successful, work on floated gyros was stopped.

The floated design was, however, used to produce the miniature rate gyros (MRGs). The accuracy of these gyros was sufficient to establish the bending of the launch vehicle which has to be taken into account in the onboard guidance scheme. MRGs are still used for this purpose, even in the PSLV. The technology for the MRGs was transferred to Hindustan Aeronautics Limited (HAL). The gyro was also found suitable for the Vijayanta tank's gun control system.

ISRO's experience with inertially stabilized platforms was used by the Tata Institute of Fundamental Research (TIFR) for the high-precision pointing of a balloon-borne infra-red telescope. It was also used in the design and development of the antenna for a ship-borne communications terminal.

The ASLV, too, used the stabilized platform system. But its closed loop guidance required inertial navigation in addition to attitude reference. So the stabilized platform system for the ASLV also incorporated three imported Ferranti accelerometers to measure the vehicle's acceleration. According to Vedachalam, ISRO began development of its own pendulous servo-accelerometers only in the early Eighties and completed this indigenous development by around 1986. The indigenous accelerometers were used only in the PSLV onwards.

In the SLV-3 and the ASLV, the equipment bay, with avionics and other electronics, was between the third and fourth stages. The equipment bay was jettisoned with the third stage. If, on the other hand, the equipment bay had been carried with the fourth stage, the injection accuracy would have improved. But the weight of the equipment bay would have greatly reduced the already small payload capability of these launch vehicles.

In the PSLV, however, guided injection was essential to achieve the orbital accuracies required and the equipment bay had to be coupled to the fourth stage. ISRO was keen to use a strapdown system to minimize the consequent loss of payload. Two stabilized platform systems, one being the primary system and the other for back-up, would have weighed about 90 kg. A strapdown system with three DTGs, in contrast, would have weighed less than 35 kg. Since each DTG could give information on two axes, the three DTGs provided ample redundancy, points out Vedachalam. If, therefore, a stabilized platform system was used, the PSLV's payload would have been reduced by about 60 kg. By this time, faster space-quality microprocessors had become available for the additional computational load of a strapdown system. So the question for the PSLV was whether strapdown-quality gyros would be available in time.

As narrated earlier, ISRO continued trying to improve the performance of its floated gyros. But, by 1980, it had also begun work on an entirely new type of gyro, the dynamically tuned gyro (DTG). While other gyro designs used fluid or gas to minimize friction, the DTG achieved the same result through a novel mechanical design. ISRO intended its DTG to be capable of being used in a strapdown system, and also be useable by both satellites and launch vehicles. The performance of the Indian DTGs is said be similar to the Teledyne gyro about which a published paper provided some limited information.

Development of the DTGs began in 1980 and the Indian DTG was ready in time to fly on the country's first operational remote sensing satellite, the IRS-1A, when it was launched in March 1988. It has since been used in all Indian remote sensing and Insat communication satellites. The Redundant Attitude Reference System (REARS) is a self-contained strapdown package for satellites, containing three DTGs. It gives information about the satellite's attitude at all times.

Only indigenous DTGs and accelerometers were used right from the first flight of the PSLV in September 1993. The Indian DTGs and accelerometers are an order of magnitude less accurate than the best sensors made elsewhere. Efforts continue to improve their performance, says Vedachalam.

IISU has initiated work on ring laser gyros which provide much greater accuracies. The ring laser gyro does not have parts rotating at high speed. Instead, it has two beams of laser light whose phases diverge depending on the rate at which the gyro is rotated. Unlike mechanical gyros, ring laser gyros are insensitive to acceleration, vibration and temperature and their accuracy is not affected by time.

The Geosynchronous Satellite Launch Vehicle (GSLV), which is expected to have its first flight in 2000, will leave the Insat satellites in a geostationary transfer orbit with a perigee of 200 km and an apogee of 36,000 km. There could be a dispersion of about 500 km in the apogee with the existing mechanical inertial sensors, according to Vedachalam. Improvement of the present inertial sensors could reduce this dispersion to 300-400 km. But to achieve the Ariane's level of dispersion of around 100 km, ring laser gyros would be needed.

Apart from gyros and accelerometers, IISU has also built and supplied momentum and reaction control wheels used in the attitude control of spacecraft. It makes the solar array drive assemblies which rotate a spacecraft's solar arrays and transfer the power generated by them to its internal power lines.

The navigation, guidance and control systems

If the propulsion system of a launch vehicle can be likened to the muscles and the inertial sensors to the eyes and the ears, then the guidance system is the 'brain' which controls the whole vehicle.

The goal that Sarabhai originally set for the SLV-3 was that it should be capable of putting a small scientific satellite into orbit with a minimum life of hundred days. To give adequate margin, this requirement was doubled to 200 days, which translated into an orbit of 400 km, says Gupta.

The *1968-69 Annual Report* of the Department of Atomic Energy stated that 'TERLS and SSTC engineers have prepared a preliminary study report for launching a modest scientific satellite of about 20-40 kg weight in 400 km orbit'. The first *Annual Report* (1972-73) of the newly formed Department of Space reiterated the goal as 'placing a scientific satellite of 40 kg in a 400 km near circular orbit'.

Achieving a circular or even near-circular orbit is not an easy task, especially for a country building a launch vehicle for the very first time. As stated at the beginning of this chapter, a satellite's orbit is determined by the velocity, height and angle of injection. This would be difficult in the SLV-3 for the following reasons:

- As in the Scout launcher on which it was modelled, the SLV-3's Inertial Measurement Unit (IMU) and guidance system was intended only to sense and correct the vehicle's attitude. No inertial guidance was possible. So if the propulsion systems, control systems or aerodynamics did not perform as expected during the flight, the uncorrected errors these introduced would affect orbital accuracy.
- Even this attitude correction would not be possible after the separation of the third stage since the equipment bay, with the avionics, would be jettisoned with it. If the equipment bay had been placed along with the fourth stage, it would have drastically reduced an already small payload capability. The result was that no guidance was available during the firing of the fourth stage and separation of the satellite.
- Since all four stages were solid, there would be more variation in their performance and, therefore, in the

> final velocity imparted to the satellite. Also, there was no way of shutting off the fourth stage's solid motor when the desired terminal velocity was reached.

All these factors became clear once the development of the launch vehicle began in earnest. It was also quickly realized that in order to ensure a minimum life of 200 days, it was only necessary to ensure that the satellite's orbit did not bring it closer than about 300 km to earth. So even an elliptical orbit became acceptable. Such an orbit could be achieved simply by ensuring that the injection took place above a certain height and with a minimum velocity. The *1975-76 Annual Report* stated that the SLV-3 would 'put 40 kg Rohini satellite into elliptical earth orbit of 300 km x 1,200 km with a minimum life of 100 days'.

The SLV-3 used what is known as 'open loop guidance' where the vehicle would try to follow a pre-programmed trajectory as faithfully as possible. According to Gupta, closed loop guidance of the sort subsequently implemented for the ASLV was not ruled out till a fairly late stage. This would have required having accelerometers as well as the onboard capability to calculate the vehicle's current position and the trajectory it should follow. In the end, the weight and size of such an inertial guidance system, and the time needed for its development, led ISRO to decide in favour of the simpler open loop guidance. The Scout showed that an open loop guidance system should work tolerably well and trajectory simulations carried out by ISRO on computers gave empirical evidence for this view.

The open loop guidance required that the optimal trajectory for the launcher be worked out in advance and programmed into the launch vehicle. In flight, the SLV-3's

IMU would immediately show any changes in the vehicle's orientation in the pitch, yaw and roll axes. No deviations would be permitted in the yaw and roll axes. But changes in the pitch axis were necessary to achieve the trajectory shown below.

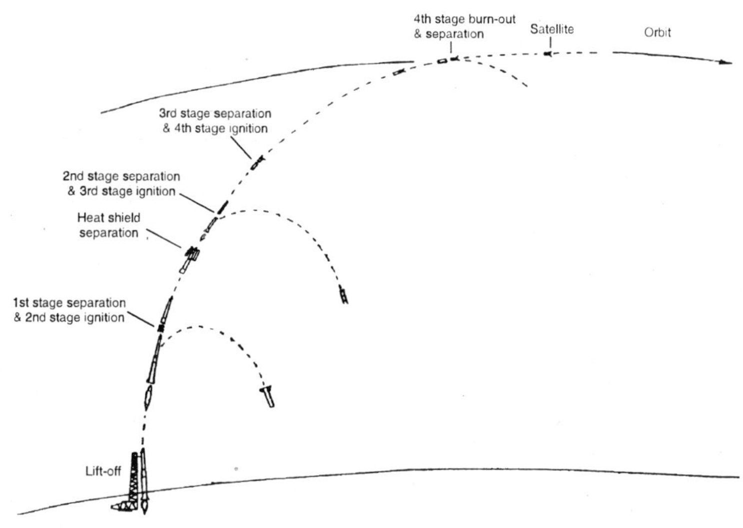

SLV-3 Trajectory

This was implemented by dividing the trajectory into a number of segments. In each segment, the vehicle's attitude in the pitch axis would be changed at a fixed predetermined rate. If the pitch of the vehicle at any instant was different from the computed value, the control systems would be activated to correct it. Operations such as separation of spent stages and ignition of the next stage would be carried out at predetermined times. It was a simple system, well-suited to the pre-microprocessor days where many different electronic components had to be soldered to put together a device. Even so, the SLV-3 guidance system did have a programmable Digital Vehicle Attitude Programmer

(DAVP). As such, the number of segments and the pitch rates for each segment could be easily changed if required.

The SLV-3 had both fin tip control as well as SITVC to control the vehicle's attitude in the pitch and yaw axes during the operation of its first stage. So the guidance system had to suitably apportion usage of the two systems. In the initial seconds of flight, till the vehicle achieved adequate velocity to make the fin tips effective, only the SITVC was used. Thereafter, to conserve the strontium perchlorate used for SITVC, fin tip control was first utilized. Only if fin tip control proved inadequate was the SITVC called upon. This method of apportioning usage of the two control plants is said to be novel.

The SLV-3 second and third stages were both equipped with thrusters. The development of these thrusters is covered in an earlier chapter. As in the Scout, the equipment bay and guidance package would be jettisoned along with the third stage. The fourth stage and satellite were therefore spin stabilized. The Scout had a turntable above the third stage which could spin up the fourth stage and the satellite before separation. This system had the advantage that the spin would reduce the impact of any disturbance during separation. ISRO, however, opted not to take this route, fearing that the turntable was too vulnerable to failure. This mechanical device had to be held down securely in flight, withstand the shocks and vibrations of flight and still work well when released.

ISRO preferred, instead, to have four small solid motor spin-up rockets which would set the fourth stage and satellite spinning immediately after separation from the third stage. After the spin-up, the fourth stage motor would be fired. Since the third stage separation occurred well above the atmosphere, at a height of over 300 km, possibilities of disturbances were considered minimal. ISRO took the view that if the spin-up rockets worked properly, there should be no problem.

Before being implemented in the SLV-3, the hardware and software for the autopilot were tested in sounding rockets. According to the annual reports, RH-125s with a roll control system — probably using just pressurized cold gas jets — were tested as early as October 1971. In June 1974, a Centaure rocket carrying miniature rate gyros, 'vehicle attitude programmes' as well as a scaled-down fibreglass heat shield were launched from Thumba. I was told that this Centaure had been fitted with a roll control system using pressurized gas. On command, the spin stabilized rocket could be despun and spun up again, clockwise and then anticlockwise.

But the most important test came in 1977-78, when a RH-560 sounding rocket was fired from Sriharikota. Instead of being spin stabilized, as sounding rockets usually are, it was three axes stabilized with a combination of SITVC, fin tip control, and monopropellant thrusters. Its onboard guidance package, which included a miniature attitude reference unit, a rate gyro package, vehicle autopilot electronics and a vehicle attitude programmer, made it possible for the rocket to follow a predetermined pitch programme. It was, in fact, a dress rehearsal for the SLV-3 and its success gave confidence in the guidance system for that launch vehicle.

The open loop system worked reasonably well during two of the SLV-3's three successful flights. The first successful flight in July 1980 — India's first successful launch — put the Rohini satellite into an elliptical orbit, 300 km by 900 km. In the second flight, in May 1981, however, the vehicle began spinning towards the end of the first stage operation. Although the rest of the flight was uneventful, the problem took its toll. There was a one per cent shortfall in velocity when the satellite was injected into orbit and the angle of injection was off by 0.5 degree. With an orbit which brought it within 183 km of the earth, it re-entered and burnt up in the atmosphere after nine days.

But for SLV-3's last flight in April 1983, the plan was to put the satellite into a 436 km by 1,021 km orbit. It was expected that there would be a dispersion of about 62 km in the perigee and 232 km in the apogee. The actual orbit achieved, 388 km by 851 km, was within these dispersion levels.

Questions have been raised about the connection in technology terms between the SLV-3 and the Scout rocket on which it was modelled. According to Gupta, only very sketchy information was available about the Scout. There was, in any case, plenty of open literature about developing the control laws required. But implementing these control methodologies required understanding the vehicle's behaviour and characteristics while flying through the air. These were specific to the launch vehicle configuration and extensive wind tunnels were necessary for this purpose. The effects of winds and other disturbances had to be calculated so that adequate control forces were available. The way the vehicle would bend and vibrate had to be taken into account in the design of the guidance system's autopilot. None of this information was readily available and had to be generated in various ways.

ASLV – From open loop to closed loop guidance

The ASLV, intended as an intermediate step between the

SLV-3 and the PSLV, was to have inertial navigation as well as 'closed loop' guidance. Instead of only correcting the vehicle's attitude, the onboard autonomous systems would be able to calculate the vehicle's exact position throughout its flight. ASLV's closed loop guidance would use this navigation information to modify the trajectory so that the satellite could be injected into orbit more accurately.

Inertial navigation is rather like dead reckoning carried out accurately to determine the launcher's exact position in space. The basic principle can be illustrated by taking the example of a car travelling due north at 20 kmph for half an hour and then due east for an hour at 30 kmph. So it has moved 10 km north and then 30 km east. It is quite easy to calculate that the car is now about 31 km away and 71 degrees north-east of the point from which it started.

If a destination is defined in terms of the starting point, then the course the car should now steer can also be calculated. Since the car is moving on the ground, it is sufficient to establish the car's position in terms of just two axes. For convenience, the geographical north-south and east-west axes have been used in this example.

In space, however, the height above the earth has also to be measured, adding a third axis. Thus, three axes, each perpendicular to the other two, are needed to fix the position of any body. Inertial navigation is carried out in terms of a three-axes inertial frame with the launch pad as its origin. If the velocity of a body along each of the three axes is known as it moves, then its position can be fixed.

As stated in the earlier section, the ASLV used a Stabilized Platform Inertial Navigation System (SPINS), with accelerometers and gyroscopes. The gyroscopes detect rotation along the three axes. But no inertial sensor can directly measure velocity. Instead, the accelerometers detected change in velocity, either acceleration or deceleration, and the onboard guidance system used this information to compute the velocity along each axis.

The stabilized platform would be carefully aligned with reference to the inertial frame before launch. During the flight, the rotation detected by the gyros is used in a stabilized platform system to counteract any change in the orientation of the platform. Irrespective of how the vehicle tilted during flight, the stabilized platform system ensured that the accelerometers always retained their original alignment. So throughout the flight, the accelerometers would give the accelerations along the three axes. The velocity and thereby the vehicle's position could be readily calculated.

The gyroscopes had another function too and that was to provide information about the launcher's attitude along the pitch, yaw and roll axes. The ASLV would be held steady in the yaw and roll axes. The onboard guidance system would suitably adjust the launch vehicle's pitch. Most launch vehicles, in fact, follow a predetermined pitch programme till they clear the dense lower atmosphere. The ASLV's closed loop guidance began only after the ignition of the second stage at a height of about 48 km. Then, till a height of about 390 km, when the third stage and the equipment bay would be jettisoned, the closed loop guidance would set the vehicle's pitch to steer it along the trajectory of its choosing.

A guidance system has essentially three major functions: navigation, guidance and control. The information from the gyros and accelerometers are used by the navigation system to compute the vehicle's attitude, its velocity along the three axes and its position.

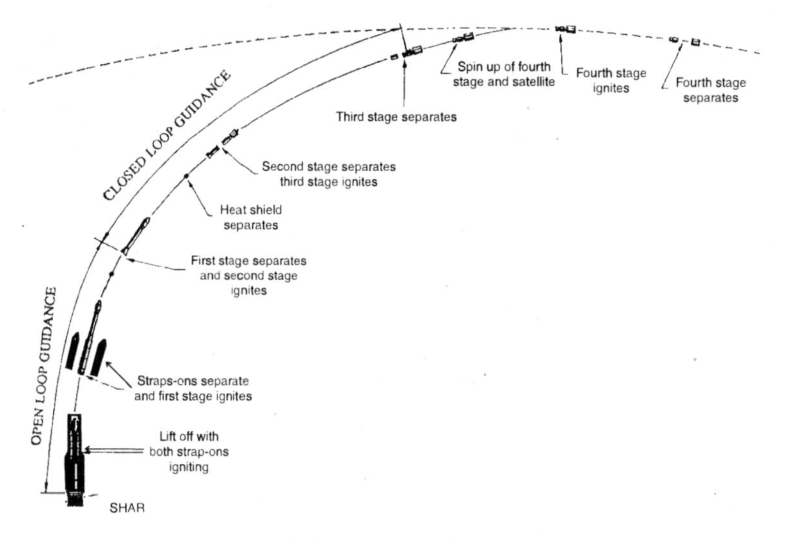

Flight Sequence of ASLV

This information is passed on to the guidance software. During the initial open loop flight through the lower atmosphere, this part of the guidance system executes the pre-programmed attitude changes. Then, in the closed loop phase, it works out the trajectory that the vehicle should follow and, on that basis, what its attitude should be. Depending on the performance of the second and third stages, the ASLV's guidance algorithms shaped its trajectory appropriately. By the end of the third stage, the launcher had to be at a point where, assuming normal performance of the fourth stage, the satellite could be injected into a 400 km near-circular orbit. By changing the trajectory to match the performance of the second and third stages as well as by choosing the most suitable time for separation and ignition of the fourth stage, the ASLV could inject a satellite with greater accuracy than the SLV-3.

The autopilot software, which performs the command function, then compares the vehicle's attitude with those coordinates required by the guidance software. It then generates suitable attitude control commands to be executed by the appropriate control power plants, such as the SITVC and thrusters.

The Motorola 6800 microprocessor powered the ASLV's onboard computers. Although there was only one stabilized platform, the data from its sensors was passed on simultaneously to two redundant chains of electronic systems for navigation, guidance and control functions. The fault-tolerant features built into the system allowed the primary chain to be shut down during flight if it was faulty. The redundant chain would then immediately take over the task.

Considering its limitations, the ASLV's closed loop system worked well. By the time of the last launch of the ASLV in May 1994, the dispersion allowed in perigee was only a third of that in the SLV-3. Since the ASLV's fourth stage too was without guidance and velocity cut-off, the dispersion in apogee remained about the same as in SLV-3. The inclination of the orbit (the angle which the orbit makes with the equator) had to be within 0.63 degree.

Injecting a satellite into a precise orbit requires more than the brute force provided by propulsion technology. Competence in control, guidance and navigation are crucial to the creation of operational launch capability. The technology for making gyros and accelerometers, which sense rotation and make the calculation of velocity possible, had to be integrated into a system involving sophisticated software and different control plants. These techniques have been mastered by very few countries.

The SLV-3, with its simpler control and guidance system, provided the initial experience which was needed. The development of suitable gyros and accelerometers,

Sarabhai discusses plans. From left: R. Aravamudan, E.V. Chitnis, M.S.R. Dev, S.C. Gupta, M.K. Mukherjee, Sarabhai, P.D. Bhavsar.

Brahm Prakash, VSSC's first director and a person revered by all, with M.R. Kurup (on the left) and Vasant Gowariker (on the right). Gowariker succeeded Brahm Prakash as VSSC director and Kurup later became director of the Shar Centre. Both played a major role in establishing solid propulsion in ISRO.

ISRO chairmen, past and present. From left: U.R. Rao, K. Kasturirangan, Satish Dhawan.

India's very first launch vehicle, SLV-3, ready for launch. Its task was to put a 40 kg Rohini satellite into an elliptical orbit. The SLV-3 put India in the select space club.

The SLV clears the launch tower, after an agonizing 36-hour countdown, on 18 July 1980. The country exploded with joy following its successful launch.

An overview of the launch complex at Sriharikota, with a PSLV carrying an IRS remote sensing satellite on the pad. The centre is the realization of Sarabhai's dream of building and launching operational satellites within the country. After the first sounding rocket was launched from Indian soil in 1963, it took three decades for the Indian rocket programme to reach this stage.

Clouds of smoke billow out as a PSLV first stage motor, one of the world's largest solid motors, is ground-tested at ISRO's Static Test and Evaluation Complex (STEX) at Sriharikota. It is probable that no other solid motor of this class was flown after just two ground tests.

The PSLV first stage motor is made up of five segments, each bearing some 25 tonnes of solid propellant. The segments are stacked right on the launch pad when the launch vehicle is integrated. Here one of the five segments is being hoisted up to be joined to the segment already in place.

One of PSLV's six solid strap-ons being hoisted to be joined to the giant first stage. Each of these strap-ons is similar to the SLV's first stage.

A.E. Muthunayagam, the man who led the liquid propulsion development in ISRO from the beginning, examines part of the PSLV second stage.

India acquired the technology for France's Viking liquid engines through a novel contract and used it to make an equivalent engine, the Vikas. Here, Vikas is being tested for its ability to swivel. Already flying in the PSLV's second stage, the Vikas engine will also be used in the GSLV's four liquid strap-ons.

The PSLV's second stage, with the Vikas engine being lifted up to be placed on top of the first stage (in the background).

A technician inspecting one half of the heat shield after its manufacture. The heat shield, made of aluminium alloy, has a diameter of 3.2 metres and is 8.3 metres long. The purpose of the shield is to protect the satellite from the heat generated as the launch vehicle passes at high speeds through the atmosphere.

In this picture, taken in the clean room in the Mobile Service Tower, the IRS satellite has already been mated to the PSLV. The two halves of the heat shield are seen on either side. The heat shield has 'acoustic blankets' inside (visible as the black material with square patterns cut into it) which reduce the high noise levels the satellite would otherwise experience. The two halves of the heat shield will be closed around the satellite to protect it. Once the atmosphere is cleared, at a height of over 100 km, the heat shield's two halves are separated and jettisoned.

The fully integrated PSLV ready for launch.

The eleven-storey-high PSLV takes off.

combined with the testing of the closed-loop guidance philosophy in the ASLV, made possible the complex and wholly indigenous guidance system for the Polar Satellite Launch Vehicle (PSLV).

CHAPTER

8

PSLV: Achieving Operational Launch Capability

As we saw in the first chapter, Sarabhai's goal, even when establishing the sounding rocket launching station at Thumba, was to ensure that India achieved self-sufficiency in launching its operational satellites. The sounding rocket programme, the SLV-3 launch vehicle and later the ASLV were the means to acquire the technological and managerial capabilities needed to build more powerful launch vehicles.

Sarabhai had foreseen that communications, direct TV broadcasting, remote sensing and meteorology would be the most important applications which satellites could provide India. These needs implied the development of two types of launch capabilities. One was to put remote sensing satellites into a polar orbit where they would circle the globe around the poles. Remote sensing satellites are often placed in a special type of polar orbit called a sun-synchronous orbit. The advantage of a polar sun-synchronous orbit is that the satellite always comes overhead a place at the same local time. With the sun always at the same angle, the lighting conditions remain comparable when images are taken on different passes.

The other would be to put communication satellites into geostationary orbit. In this orbit, some 36,000 km above the equator, the satellite matches the earth's rotation and therefore appears stationary when seen from the ground. Since antennas on the ground do not have to track a moving satellite, they can be kept simple. The satellite can act as a space-based relay station, passing on radio signals between distant places. From its vantage point high above, the satellite can also broadcast TV signals over a wide area. So satellites for communications and direct broadcasting are usually placed in the geostationary orbit. India's Insat satellites often carry meteorological cameras as well. From

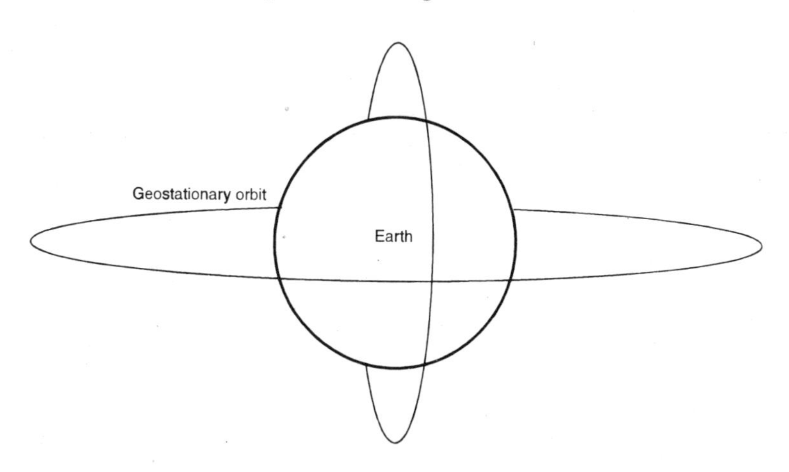

Polar Sun-synchronous Orbit

geostationary orbit, these cameras can constantly monitor fast changing weather patterns, including cyclones.

Instead of putting these satellites directly into geostationary orbit, launch vehicles usually leave them in an elliptical orbit, called the geostationary transfer orbit (GTO). The satellites are equipped with solid or liquid rocket motors which are fired to take them from the transfer orbit to the final geostationary orbit.

Much more powerful launchers are required to put satellites into the GTO than those needed to put satellites

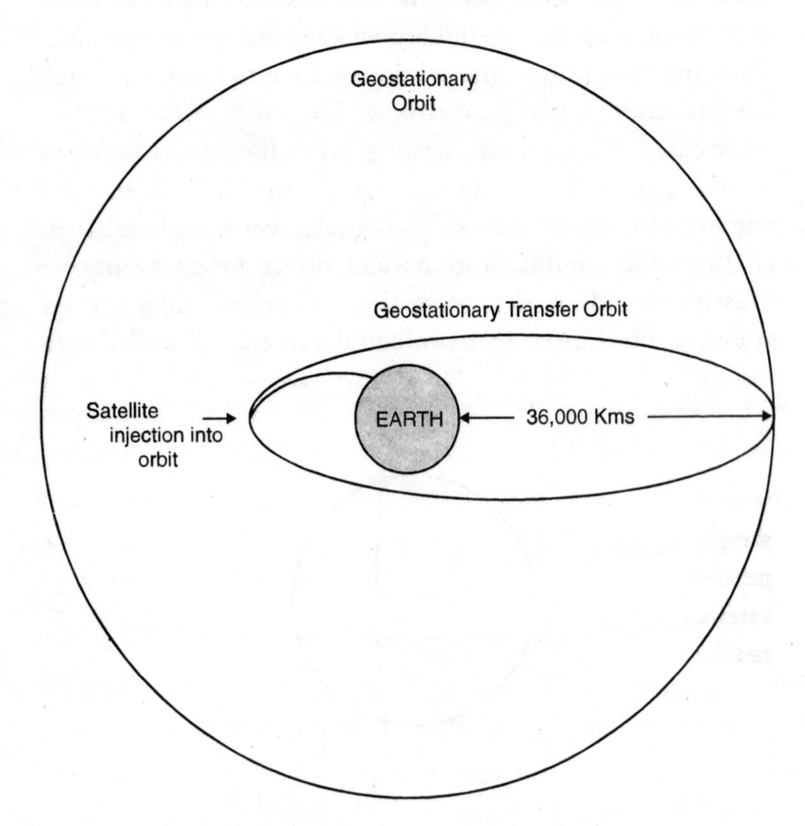

Geostationary Transfer Orbit to Geostationary Orbit

of equivalent weight into polar sun-synchronous orbit. The difference in the capabilities required can be gauged by the fact that the current Polar Satellite Launch Vehicle (PSLV), which can put a 1,200 kg satellite into an 800 km polar orbit, would be able to take just 800 kg into the GTO.

Sarabhai planned to go straight from the SLV-3 to a launcher for putting communication satellites into geostationary orbit. It is quite likely that he took the view that once GTO launch capability was achieved, it would also imply that heavier satellites could be put into polar orbits.

The *Atomic Energy and Space Research: A Profile for the Decade 1970-80* had stated:

> SLV-3 would be followed in the period 1975-79 by satellite launch vehicles using more powerful motors and it is the objective of the Space Science & Technology Centre to develop by the end of the 1970s a launch vehicle capable of putting a 500 kg satellite into synchronous orbit at 40,000 km. This is the type of capability which is needed to fully exploit, on our own, the vast potential arising from the practical applications of space science and technology.

As the *Profile* suggests later on, such a satellite would weigh about 1,200 kg at launch. A substantial part of the weight of such a satellite would be taken up by propellants needed to take it from GTO to geostationary orbit. The satellite would therefore weigh only about 500 kg when it reached geostationary orbit.

With the wisdom of hindsight, it is clear that Sarabhai underestimated the difficulties of launcher development. The *Profile* had been prepared in 1970 when even the design of SLV-3 had not been completed. It would be impossible to have the SLV-3 ready by 1974 and the bigger launcher by the end of the Seventies. The sheer magnitude of the latter task is shown by the fact that it is only now, three decades later, that ISRO is in a position to put a satellite of that class into the GTO.

Within a couple of months of Dhawan taking over as head of the space programme in September 1972, a study group was set up to study future launch vehicle configurations. The very first *Annual Report* of the newly formed Department of Space, covering the 1972-73 period, stated that 'a team has been constituted to analyse, study and report on an optimum approach to the development of a multistage launch vehicle, capable of placing a satellite of about 800 kg weight in synchronous equatorial orbit'. Once again, the emphasis was on developing a launch vehicle to put a satellite into geostationary orbit. This time, though, the satellite would weigh 800 kg when it reached that orbit.

In August 1972, ISRO had convened a national seminar on the Indian Programme for Space Research Applications at Ahmedabad. The seminar was probably intended as a way of building up support among possible user agencies for an Indian National Satellite (Insat), a communications satellite. It was after this seminar, in November 1972, that a study group was set up to examine configurations for launchers which could put such a satellite into orbit.

According to R.M. Vasagam, the man who headed the team, they studied the configurations of existing launchers before submitting a report around August 1973. Vasagam says that the configuration they recommended involved a cluster of four liquid engines, each producing a thrust of 60 tonnes, for the first stage. A similar engine would be used for the second stage. The third stage would have two cryogenic engines, each producing 7.5 tonnes thrust. The launcher had a fourth stage with a pressure-fed liquid engine to take the satellite from transfer orbit to geostationary orbit. The first stage would be surrounded by four solid strap-on motors weighing about 10 tonnes each. Since subsequent annual reports mention further studies being conducted, it is possible that the configurations were refined some more.

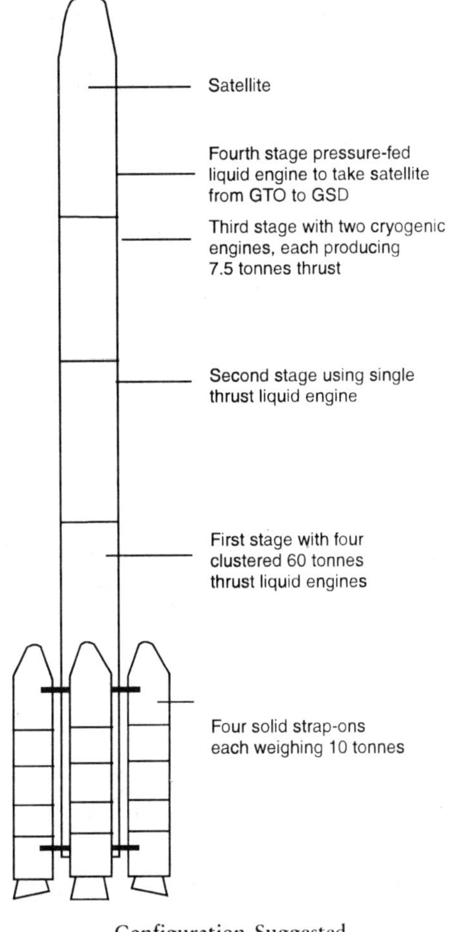

Configuration Suggested by Vasagam Committee

Change in focus

By the late Seventies, however, there was an important shift away from building a launch vehicle for putting communication satellites into geostationary orbit towards first developing a launcher to carry indigenous remote sensing satellites. ISRO was well into the SLV-3 project by this time. Having encountered the difficulties in developing even such a simple launcher, ISRO would have had a better appreciation of the problems involved in building more powerful launchers.

Moreover, after Sarabhai's death, the concept and need for Insat communication satellites did not receive much support from key user-ministries such as the Ministry of Communications. Indeed, the proposal for Insat-1 series of satellites, which would be built abroad, was cleared by the government only in early 1981, after ISRO secured the backing of the Ministry of Communications, the Ministry of Tourism & Civil Aviation and the Ministry of Information & Broadcasting. Clearance for the initial Insat-2 satellites, which would be built indigenously, was given only in mid-1985. Without firm commitments on the Insat satellites, the case for a launch vehicle to carry such satellites became difficult to sustain in the late Seventies.

As it turned out, developing the PSLV itself posed enough challenges. A launcher to put communications satellites into orbit could have taken twelve to fifteen years, or even more, to develop. It would also have required a much larger financial commitment. Without the backing of user-ministries for an Insat type of satellite, it would have been difficult to justify such an expensive and time-consuming project to the government. It therefore appeared sensible to develop the PSLV first and then use that experience to move onto the Geosynchronous Satellite Launch Vehicle (GSLV).

In December 1977, the VSSC director, Brahm Prakash, issued an office order establishing a committee headed by S. Srinivasan to recommend configurations for a launch vehicle which would put a 600 kg class remote sensing satellite into a 550 km orbit. Srinivasan later became the PSLV project director. Subsequently, he was director of Shar (the Sriharikota centre) and then of VSSC.

The guidelines given by Brahm Prakash to the Srinivasan committee for the PSLV configuration said that 'future growth possibilities should not be ignored but neither should these be made over-riding'. The objective was to build a launch vehicle which would give the lowest cost per kilogramme in orbit. Although the development cost would not be amortized over a number of launches, unit vehicle costs and development costs had to be minimized.

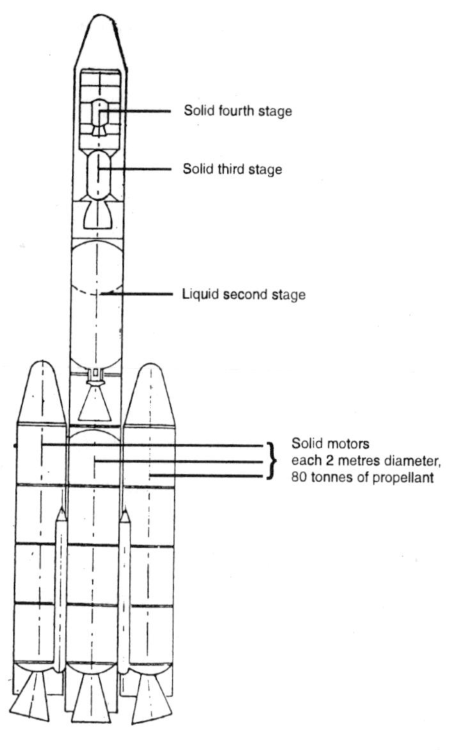

An Early PSLV Configuration

The Srinivasan Committee studied thirty-five configurations before submitting its report in April 1978. The report suggested four configurations. The preferred configuration involved a large solid motor, 2 metres in diameter and carrying 48 tonnes of propellant. One such motor would form the first stage, with two similar motors as strap-ons. The second stage would have a liquid engine and the upper two stages would both be solid. These initial configurations were followed up with further studies. By the time *Countdown* carried an article on the PSLV in its November 1980 issue, the big solids had doubled in size. Instead of 48 tonnes, they would each carry 80 tonnes of propellant.

A major bone of contention which emerged between the launch vehicle and satellite development groups concerned the payload capabilities of the launcher. The present ISRO chairman, K. Kasturirangan, who was project director for the first IRS satellite, remembers those days. The payload capability of 600 kg was inadequate for the kind of sensors which the remote sensing satellite had to carry, he says. In the end, the payload requirements for the PSLV were raised. By the time the PSLV Project Report was submitted by Srinivasan in December 1981, it had been decided that the launch vehicle would have to put a minimum of 1,000 kg in 900 km polar orbit.

The four stage launch vehicle configuration which emerged as a result of this enhanced requirement needed development of one of the world's largest solid motors, 2.8 metres in diameter holding 125 tonnes of propellant. This giant first stage would be surrounded by six solid strap-ons, each of them similar to the first stage of SLV-3 and ASLV. The second stage of this operational launcher would be built around the Indian equivalent of the Viking engine. The upper two stages would both be solid. It was this configuration which was given the go-ahead by the government in June 1982. It was estimated to cost Rs 311.57 crore and take five years to develop.

About a year later, an important change was made, with the solid fourth stage being substituted by a liquid stage. This change was considered necessary since the accuracy with which the IRS satellites had to be put into orbit — within 15 km in terms of orbital height and within 0.1 degree of the desired orbital inclination — could not be achieved with a solid stage. A liquid stage, with the ability to shut off the engine when needed, would be able to take care of situations when the lower stages either over-

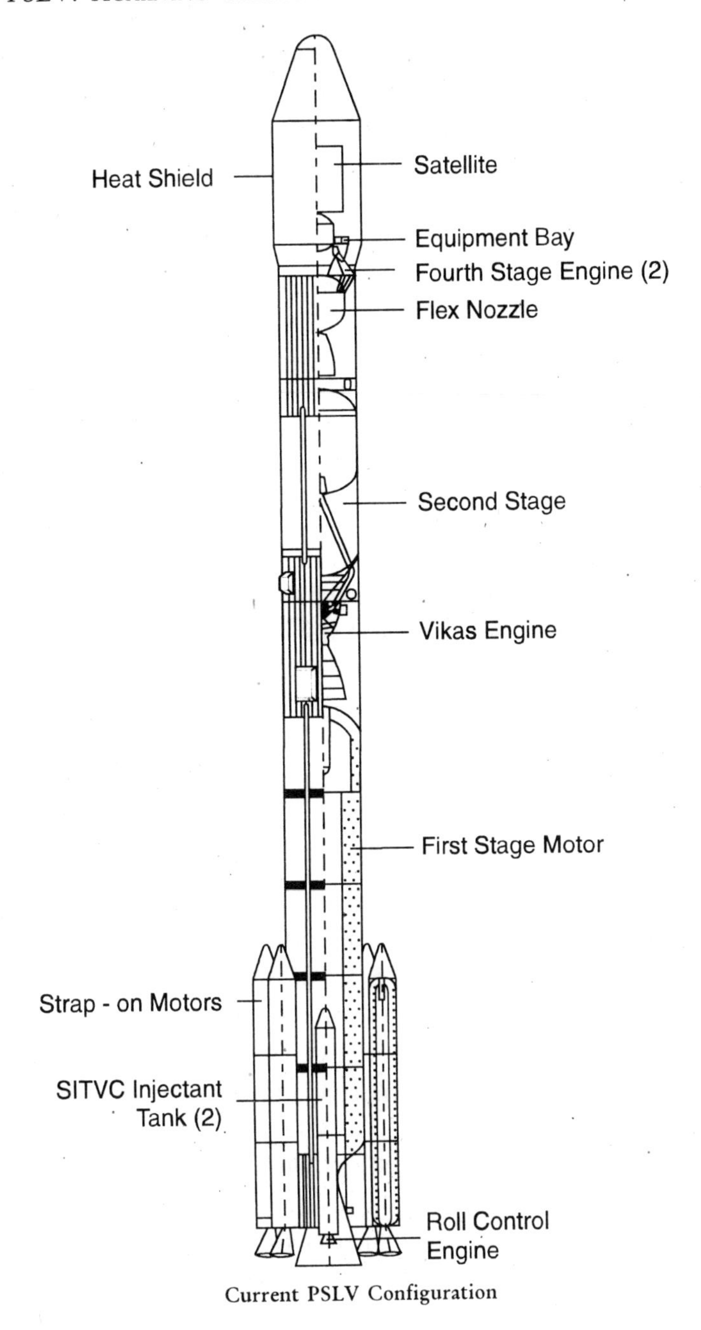

Current PSLV Configuration

performed or underperformed. Solid stages, on the other hand, operate till their propellant is exhausted and, even with the most stringent manufacturing conditions, there would still be variations in their performance.

Two more changes were made in the configuration. The propellant loading in the first stage was increased to 129 tonnes. The diameter of the liquid second stage was increased from 2.6 metres to 2.8 metres so that the core of the launcher had a uniform diameter right through, except for the heat shield. Consequently, the second stage would carry 4 tonnes more propellant.

The solid–liquid debate

The PSLV was ISRO's first operational launch vehicle and would become the basis for its next launch vehicle, the Geosynchronous Satellite Launch Vehicle (GSLV). So it is important to understand why ISRO chose solid propulsion, instead of liquid engines, for the PSLV's first stage. If the United States, Europe and Japan chose liquid propulsion for the core stages of their medium and heavy launchers, it was not because they lacked solid propulsion technology. Not only are liquid engines more energy-efficient than solids, but the modular nature of liquid engines can be exploited to advantage. Several liquid engines can be clustered together to give the thrust needed. As the weight of satellites to be launched increases, solids rockets become disproportionately more massive and liquid engines become increasingly attractive.

Europe's highly successful Ariane-4 launcher uses a single Viking engine for the second stage and four similar engines clustered together in the first stage to provide the thrust needed at lift-off. A similar configuration had been suggested by the Vasagam Committee. Global experience has been that once fabrication, assembly and testing protocols are put in place, liquid engines and stages can be turned out in large numbers without compromising their reliability. Each Ariane-4 launcher uses anywhere from five to nine Viking engines (depending on how many liquid strap-ons are used) and the 1,000th Viking engine were delivered in June 1999. That is nothing compared to the Russians, masters of liquid engine technology. Their RD-107 is the world's most-used rocket engine, thousands of which have been built and launched.

India stands out among countries with launch capabilities as the only country not to have developed the technology for clustering liquid engines. Since the contract with SEP gave ISRO the Viking technology on a platter, why did it opt out of clustering these engines for the PSLV first stage and decide to build a huge solid motor instead?

The answer seems to be that at the time when the PSLV configurations were being studied and finalized, ISRO was still not confident of its liquid propulsion technology. In the late Seventies and early Eighties when the PSLV was configured, the 3 tonne engine was the most powerful liquid engine which ISRO had actually built. Moreover this engine was only pressure-fed. Pressure feeding is simpler, but suitable only for liquid engines of lower thrust. The more complicated turbopump system is needed for high-thrust liquid engines.

The Viking, which in its Indian version would be called 'Vikas', produces about 74 tonnes of thrust. This turbopump-fed engine would be the most complicated liquid propulsion development undertaken by ISRO up to that time. ISRO took the view that getting this engine made in India and building a stage with one engine would itself be quite difficult. Clustering was, after all, a technology with its own problems which would have to be overcome. The organization's capabilities in solid propulsion, on the other hand, were better and proven. So going from the 10 tonne motor of the SLV-3 and ASLV first stage to the 125 tonne (later 129 tonne) first stage of the PSLV appeared much less risky than clustering of a still-to-be-built engine.

Subsequently, after the Vikas had been successfully tested and when configurations for the GSLV were being considered, ISRO's liquid propulsion engineers did put forward a strong case for clustering the Vikas engine. But, by that time, using the solid boosters being developed for the PSLV was judged to be quicker and cheaper than developing a new stage with such a configuration.

Even without undertaking the clustering of the Vikas engines, each one of the four stages of the PSLV posed its own technological problems. As a result, the PSLV was delayed by five years. The cost of the project went up by over Rs 100 crore, an increase of about 33 per cent, by the time it flew in September 1993.

Solid propulsion

Solid propulsion in ISRO had grown in the Seventies from outdated technology suitable only for sounding rockets to stages with modern high-energy propellants for the SLV-3. The capability had been created to produce resins and other chemicals indigenously, to develop propellant formulations, make large solid motors and also to test them. With the PSLV, ISRO's solid propulsion groups set out to prove that they were as good as the best in the world.

The PSLV's first stage would be one of the largest solid motors in the world. Its third stage would, similarly, be one of the biggest upper stage solid motors. Such development does not come easily and many obstacles had to be overcome before both stages were ready for flight. Both stages have worked well in flight. ISRO has further improved their performance and thereby increased the payload which the PSLV can carry.

The segmented first stage of the SLV-3 has become ISRO's most flown solid motor. The ASLV used three such motors, one for the first stage and two as strap-ons. The PSLV has six of them as strap-ons. In all, as of December 1999, forty-eight of these motors have flown. Only once, when the first stage failed to ignite in the first flight of ASLV, has there been a problem with any of them.

At the time of the ASLV, segments for both strap-ons were cast simultaneously, so that the thrust they produced would be closely matched. But, by the time of the PSLV, ISRO could cast the segments separately and still be confident that performance of all six strap-ons would be very similar.

Choosing HTPB resin

One of the important decisions made in the PSLV was to base all its solid propellants on HTPB (hydroxyl terminated polybutadiene). In the SLV-3, and later the ASLV, the lower stages used solid propellants based on PBAN and the upper stages had HEF-20-based solid propellants. When possible configurations for the PSLV were being considered, the issue of which resin to use for its solid motors came up.

A technical committee (of which Gowariker was a member) set up in the late Seventies by the American Institute of Aeronautics and Astronautics (AIAA) to study resin choices for solid motors, had indicated that HTPB would be the workhorse resin for solid rockets in the years ahead.

This did not, however, make HTPB the automatic choice for the PSLV solid motors. Apart from PBAN, the indigenously-developed ISRO polyol was also in the running. Solid propellant formulations were made from all three resins and tested for a variety of properties. Formulations using HTPB gave slightly greater energy than the other two resins. HTPB formulations were also able to meet the mechanical strength specifications, which neither PBAN or ISRO polyol-based formulations could. Although US propellant manufacturers had achieved good results with PBAN — it is, for instance, used in the space shuttle's Solid Rocket Boosters, the world's largest solid motors — it had not been possible to reproduce those results in India, according to Rajaram Nagappa, who was involved in the selection of HTPB and was till recently associate director at VSSC.

As a result, ISRO became one of the earliest to switch to HTPB for large solid motors. HTPB is now used in the recent Solid Rocket Motor Upgrade (SRMU) for the US launcher Titan IVB rocket, the two solid boosters for Europe's latest Ariane-5 launcher and the Japanese H2's boosters as well.

Imported HTPB had been utilized for the early tests to decide which propellant resin should be used for the PSLV's solid motors. But, by the late Seventies, VSSC had already started working on a chemical process for synthesizing HTPB. Transportation of large enough quantities of butadiene gas was no longer the problem it had once been. The National Organics Chemical Industries (Nocil) had established its petrochemical plant near Bombay and the butadiene gas could be trucked down in tankers to Trivandrum. The *1977-78 Annual Report* reported that a benchscale process had been established. Two years later, the *Annual Report* said that the process had been scaled up to 35 kg per batch. The technology for HTPB production was transferred to Nocil which had agreed to carry out the industrial production of HTPB.

The first stage of PSLV

With a diameter of 2.8 metres (close to 10 feet), PSLV's first stage was so wide that even the tallest human could stand upright and walk through its casing. It would hold 129 tonnes of solid propellant, making it one of the world's largest solid motors. It would be two times longer, three times wider and more than twelve times heavier than the first stage of the SLV-3. The giant solid motor was designated the S-125, the '125' deriving from its initial propellant loading of 125 tonnes.

A PSLV first stage motor being carefully checked

It would be made of five segments, each carrying some 25 tonnes of propellant. Success with the SLV-3 first stage, which too had been segmented, had provided the experience and confidence to attempt such large segments. But each PSLV first stage segment would be two and a half times heavier than the entire SLV-3 first stage. The difficulties of making a solid motor increase with its size and even with the SLV-3 experience behind them, the development of the PSLV first stage was not something that ISRO and its solid propulsion group could take for granted.

Before the PSLV project was formally approved, ISRO began the expansion of the Solid Propellant Rocket Booster (SPROB) plant so that segments of this size could be cast. The Static Test and Evaluation Complex (STEX) was augmented for testing the PSLV's solid motors.

15 CDV 6 or maraging steel?

If the decision to use HTPB did not create any major ripples, the issue of whether maraging steel should be used for the casing of the first stage aroused heated debate and acrimony.

A VSSC internal committee is said to have initially recommended that the first stage be made of 15 CDV 6. The French had specified this steel for their Centaure sounding rockets which India began licence producing from 1969. Subsequently, 15 CDV 6 had been used for the lower stage motor casings in both the SLV-3 and ASLV. Since its properties and fabrication methods were well understood by ISRO and industry, the case for continuing with 15 CDV 6 for the PSLV first stage motor was strong. This conservative view held that casting and assembling the huge first stage posed enough challenges without adding more by using new materials like maraging steel.

But others felt that it would be stupid to persist with a material like 15 CDV 6 for such a large motor since the weight penalty would be substantial. They argued that the first stage casing should be made of much stronger steel. It would then be possible to make the walls of the casing thinner, without compromising the casing's ability to withstand the high pressures which build up when the solid propellant is burning inside. A lighter motor casing would improve the motor's performance in flight and contribute to a higher payload in orbit.

The problem is that increase in strength is usually accompanied by a decrease in the metal's 'fracture toughness'. Under stress, a material with lower fracture toughness is more liable to allow flaws, which are inevitably present, to grow and propagate rapidly, leading to catastrophic failures. It is hardly a desirable trait in a material for casing subjected to high pressure.

The issue was whether it was possible to use a metal which combined high strength with sufficient fracture toughness. A candidate material which could meet this requirement was maraging steel. This was an ultra-high-strength steel developed in the late Fifties. Maraging steel could be twice as strong as 15 CDV 6 with comparable fracture toughness. The casing for the PSLV first stage motor has to be able to withstand pressures about fifty-three times higher than normal atmospheric pressure. If maraging steel was used, the casing need be only 8 mm thick. It would have to be 14 mm thick if 15 CDV 6 was used. So a maraging steel case would have only about half the weight of a casing made of 15 CDV 6.

The annual reports have references to work on maraging steel being carried out at VSSC from the mid-Seventies. These capabilities were, however, nowhere near what was needed to produce the PSLV casings and considerable problems remained. Maraging steel had strategic uses and its import would be difficult, if not impossible, as ISRO discovered when it tried to buy some while the technology was being developed. In any case, ISRO could not afford to become dependent on imports of a material needed in large quantities and which could be embargoed at any time. So production of the steel on an industrial scale within the country had to be established. In addition, a heat treatment process needed to be developed in order to achieve the strength and fracture toughness required.

The controversy over maraging steel reached such proportions that in 1981 Dhawan appointed a national committee headed by Brahm Prakash (who had by then retired from the directorship of VSSC). Apart from 15 CDV 6 and maraging steel, the committee examined the possibility of using another material, D6 AC. The committee unanimously recommended using maraging steel.

In a paper on ISRO's solid motors presented by VSSC scientists at the International Astronautical Federation meeting at Bangalore in 1988, the reason for this choice has been discussed at length. Use of 15 CDV 6 would make the motor case heavy and lead to loss in payload. While D6 AC could offer lower cost than maraging steel, there were problems with welding this material. Setting up facilities for weld-free construction would be costly and time-consuming. Except for cost, maraging steel satisfied all the criteria for the first stage material. The paper pointed out, however, that if material, fabrication, heat treatment and facility costs were all taken into account, the overall cost difference was not too high for maraging steel.

Many in ISRO still believe that the choice was also motivated by the needs of the atomic energy programme.

Maraging steel has been used to make high-speed centrifuges for uranium enrichment. This sounds somewhat improbable since the Indian atomic energy programme has outstanding and proven metallurgical capability. It is unlikely to look to space for a helping hand.

At any rate, after the recommendation of the Brahm Prakash Committee, ISRO set up a Maraging Steel Indigenization Programme (MSIP) in 1982. The programme was successfully completed by the end of 1983 as a result of a unique partnership between VSSC, the Hyderabad-based public sector speciality metals manufacturer, Mishra Dhatu Nigam Limited (Midhani), the Defence Metallurgical Research Laboratory (DMRL) and the Rourkela Steel Plant.

Making the giant first stage

In 1985-86, a SLV-3 third stage motor was cast with HTPB-based propellant and successfully ground-tested. A year later, two segments of the SLV-3 first stage were made and then put together to demonstrate the feasibility of the new 'dry joining' technique which would be used to assemble the PSLV first stage's five segments.

In May 1988, a PSLV first stage segment was cast for the first time. It took sixty-four hours at a stretch to carry out this operation. The maraging steel case had come from Larsen & Toubro, the HTPB from Nocil and the ammonium perchlorate from ISRO's own plant at Alwaye.

A year later, on 21 October 1989, the giant first stage motor was successfully ground-tested at STEX. The performance of the motor matched the predicted values very closely. It was a big relief and, as an article in the ISRO publication *Space India* remarked, a shot in the arm for the PSLV project. A second test was carried out in March 1991. This too was a complete success.

Each such test is expensive as the motor costs something like Rs 10 crore. Since the results from both tests matched closely and also agreed with the predicted performance, it was decided that two ground tests were adequate for going ahead with flight versions of the first stage. The PSLV first stage is almost certainly the only large solid motor anywhere in the world to have been flown after so few ground tests. Solid motors are unforgiving with even the smallest flaw, as the space shuttle Challenger disaster demonstrated. If the PSLV first stage exploded on or near the launch pad, it would have wrecked the pad and delayed the launcher programme by at least a few years.

Improving the first stage

There has since been one more ground test when the first stage was improved to increase the PSLV's payload by about 80 kg. Nearly 9 tonnes of propellant was added to the first stage. Insulation and other inert weight in the stage was reduced. The propellant formulation was modified to ease processing. The S-139 motor, as it is called, was successfully ground-tested in April 1997 and the PSLV with the new motor flew in September that year.

In an effort to stem the rising cost of maraging steel without sacrificing its beneficial properties, VSSC's materials group has developed a cobalt-free low-nickel variety of maraging steel which it says could be some 37 per cent cheaper. Maraging steel, which cost Rs 700-800 per kg in the mid-Eighties, was reportedly now being supplied by Midhani at between Rs 2,000-3,000 per kg. The rising cost of nickel and cobalt, which have to be imported, was said to be a major reason.

But it is unlikely that ISRO will want to change the casing for the present first stage motor. Such a switch would be expensive. Every step of the production process has to be proven and in all probability at least one ground test of the full motor with the new casing would be needed. Besides, space is a conservative business, 'if it ain't broke, don't fix it' being the ruling tenet the world over. ISRO, however, does plan to build still bigger solid motors and the use of this new steel would be considered then.

The solid third stage

Just as the first stage was one of the world's largest solid motors, PSLV's third stage would be one of the biggest upper stage solid motors. It would be the first ISRO motor to have a flex nozzle which could be swivelled. By swivelling the nozzle, the direction of the thrust could be changed in order to correct or change the vehicle's orientation along two axes (pitch and yaw).

The PSLV third stage after being integrated with the fourth stage

In order to reduce inert weight so that the performance of the motor in flight would improve, the third stage motor casing would be made of Kevlar, the high-strength fibre. Although the SLV-3's fourth stage motor, which was used in the ASLV too, was also made of Kevlar, the third stage motor casing posed considerable headaches. Unlike in the SLV-3 and ASLV motors, the third stage motor had unequal-sized openings at the top and bottom. The opening at the bottom had to be large to permit swivelling of the nozzle. This opening would be nearly twice the size of the opening at the top where the igniter to fire the motor would be fitted.

The unequal openings made winding of the Kevlar fibre difficult. The motor case had to be capable of withstanding pressures sixty-three times higher than normal atmospheric pressure. This was more than twice the pressure which ASLV's fourth stage Kevlar case was designed to take. When the PSLV third stage motor casing was tested under pressure, the Kevlar fibres slipped. It was 1988 before a winding method was discovered whereby the Kevlar fibres stayed in place under pressure.

The motor case was filled with an HTPB-based propellant in December that year. In April 1989, the third stage motor was successfully fired for the first time. In January 1990, a second test followed, this time with the 'flex seal' mechanism, which allows the nozzle to swivel, in place. For this test, however, the nozzle was not allowed to swivel. The test was a success and, after the initial problems with the case, the third stage development appeared to be going smoothly.

Then, as so often happens in space matters, trouble struck. In the third test later that year and the subsequent one, the motor failed when the nozzle was swivelled. Silicone-based thermal protection for the flex seal was developed and other design changes made. To the relief of

ISRO's engineers, the fifth and sixth tests in 1991 were flawless. The seventh test was successfully carried out at the high-altitude test facility which simulates the vacuum which the motor would encounter during actual flight. There were two more tests, the ninth and last being performed in December 1992. The third stage was now ready to fly in the PSLV.

The solid third stage has performed well in flight. ISRO has already improved the performance of this motor. Inert weight, principally of insulation inside the motor case, has been considerably reduced and the weight of the propellant increased. The improved motor flew in the third flight of the PSLV in March 1996 and is estimated to have increased the payload it can carry into polar sun-synchronous orbit by about 50 kg.

A high-performance version is currently under development. The improvements include making the openings at the top and bottom of the motor case more equal in size. As a result, the fibre winding can be made simpler and less heavy. The high-performance motor is likely to be ground-tested in 2000. Its use will increase the PSLV's payload capability in polar sun-synchronous orbit by about 90 kg.

Growth in solid propulsion technology

It is not just that the PSLV's first and third stage motors are among the largest of their class in the world. Quite apart from their size, the PSLV solid motors represent a considerable advance in technology over their counterparts in the SLV-3 and ASLV programmes. Moreover, both the first and third stages have been improved during the course of the PSLV programme.

The performance of any rocket stage, irrespective of the propulsion system it uses, is determined by two factors. One is the amount of energy which can be extracted from the propellants. Rocket engineers call this specific impulse. The specific impulse gives the amount of thrust which will be produced when a given quantity of propellant is consumed every second. The other factor which determines performance is the ratio of the propellant weight to the total stage weight. This is called the mass fraction. The lower the inert weight of the stage, the greater will be the mass fraction and better the performance of the stage.

Graph I shows the improvement in performance of the ISRO first and third stage solid motors, from the SLV-3 and ASLV programmes to the advances made during the PSLV programme. For purposes of comparison, the performances of some of the well-known lower and upper solid stages of foreign rockets have also been included.

The terminal velocity which would be reached if that stage were fired alone has been taken as the index of performance. The velocity thus reached is determined by the specific impulse and the mass fraction. As pointed out before, both these factors are influenced by the level of technology reached. It also allows stages of different sizes to be compared.

The graph shows the dramatic improvement in the third stage performance during the PSLV programme. When the new high-performance third stage is ready, ISRO's technology for solid upper stages will be on par with world standards. The mass ratio and the specific impulse of the high-performance third stage will compare well with that of the Orbus 21D used as the third stage of the US Athena-II launch vehicle.

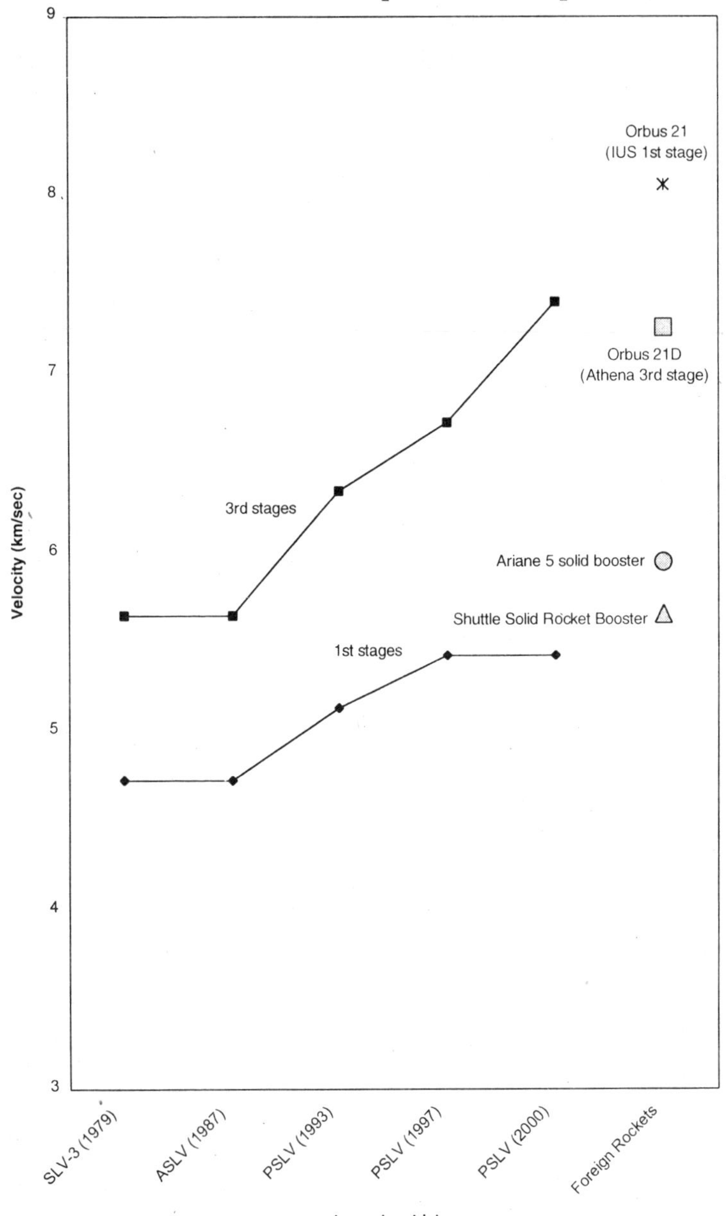
Growth in Solid Propulsion - Graph I
Velocity (km/sec)
9
8
7
6
5
4
3
Orbus 21
(IUS 1st stage)
Orbus 21D
(Athena 3rd stage)
3rd stages
Ariane 5 solid booster
Shuttle Solid Rocket Booster
1st stages
SLV-3 (1979)
ASLV (1987)
PSLV (1993)
PSLV (1997)
PSLV (2000)
Foreign Rockets
Launch vehicles

But the PSLV's high-performance third stage does not match the performance of the Orbus 21 solid motor which forms the first stage of the Inertial Upper Stage (IUS). The greater weight of the PSLV third stage nozzle appears to have lowered its performance. The Orbus 21, on the other hand, uses a lightweight carbon nozzle.

Improvements to the solid first stages have been less dramatic. The performance of the PSLV first stage is much better than that of the SLV-3 and ASLV. In addition, the S-139 gave a better performance than the S-125 which flew as the PSLV first stage in its initial flights. Even so, the S-139 does not match the performance of the shuttle's Solid Rocket Booster or that of the solid boosters of Ariane-5. But ISRO has not had much incentive to further improve the first stage since the benefit in terms of increased payload is outweighed by the high costs of ground tests.

Liquid propulsion

In the SLV-3 and the ASLV, solid propulsion was the undisputed star of the show. Liquid propellant engines had only a small supporting role, providing thrusters to correct the vehicle's orientation. But liquid propulsion moved to centre stage in the PSLV, powering two of its stages.

The second stage would be built around the Viking engine technology from France. The Indian equivalent, Vikas, would be the most powerful liquid engine built by ISRO, generating some 74 tonnes of thrust. It would also be the most complex liquid engine India had made, using a gas generator and a turbopump to drive propellants from the tanks to the thrust chamber where they would burn. Even today, all of ISRO's other liquid engines are pressure-fed systems. The Vikas will remain ISRO's sole turbopump driven engine till an indigenous cyrogenic engine is successfully developed.

The indigenous experience in building pressure-fed liquid engines was also utilized in the PSLV. The fourth stage of the PSLV was built around two such engines. These are the only ISRO-made operational engines which use regenerative cooling. In regenerative cooling, one of the propellants is passed around the combustion chamber to cool it down.

PSLV's second stage

By 1978-79, most of the ISRO team had returned from France with the technology for the Viking engine. There was reportedly considerable hostility towards the idea of using the engine for any of the PSLV stages. 'If the majority in VSSC had their way, the PSLV would have been an all-solid launch vehicle, like the SLV-3 and the ASLV,' remarks one liquid propulsion engineer.

The Vikas engine became a reality only because of the overall leadership of Brahm Prakash who headed the board which oversaw the acquisition of the Viking technology, says Y.S. Rajan, then ISRO's scientific secretary. The guidelines given by Brahm Prakash to the Srinivasan Committee asked them to keep in mind 'utilisation of liquid propulsion systems, particularly the experience built up so far in ISRO and the national agencies' when examining possible configurations for the PSLV.

When the Srinivasan Committee submitted its study report in mid-1978, three of the four configurations suggested had at least one liquid stage. The report speaks about the acquisition of the Viking engine technology and the possibility of making the engine and stage in India within about four years.

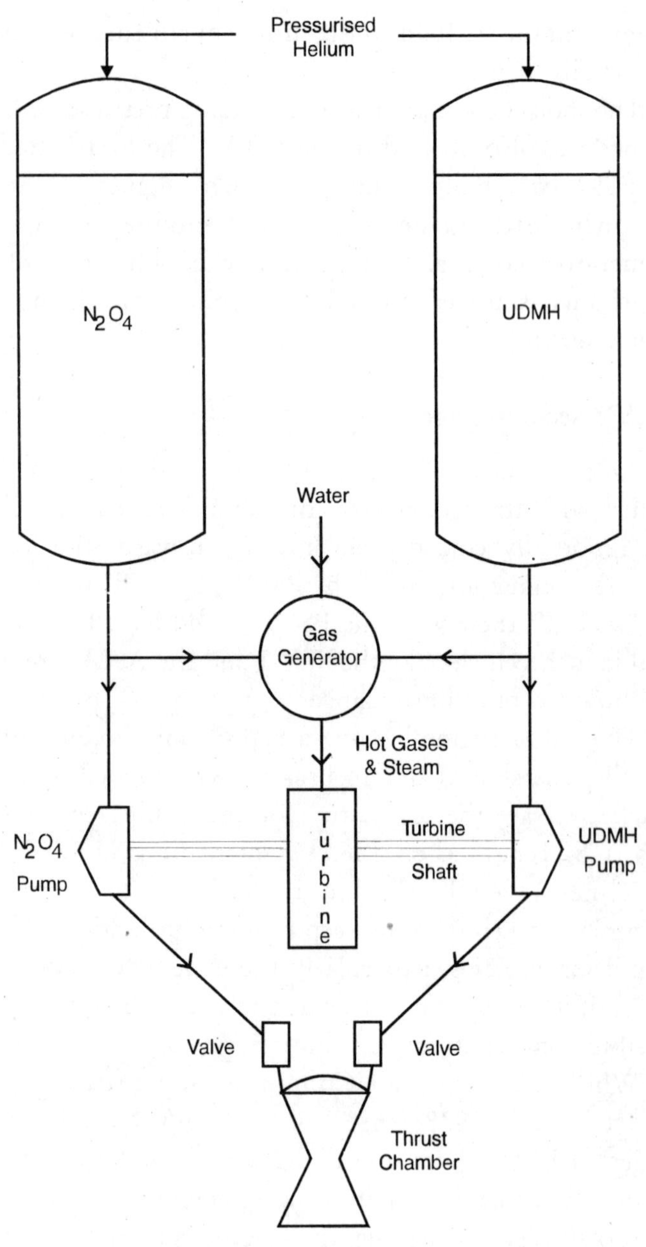

Simplified Schematic of Stage with Vikas Engine

Nevertheless, the warfare over the use of the Vikas in the PSLV seems to have continued. One issue raised was whether the technology had been really acquired and whether the engine could be successfully fabricated in India. Another issue was whether a suitable stage could be built using this engine. The stage, as will be related later, is just as complex as the engine. The contract with SEP was only for the engine. The engineers who went to France retorted that they had picked up everything needed to build the stage as well. It was only around 1981-82, about the time when the PSLV project itself was cleared, that fabrication of the Vikas engine finally began.

Likewise, the search for a site where liquid engines could be tested began soon after the Viking deal with France was signed. Around a dozen sites in various parts of the country were evaluated in exhaustive detail. The final decision on locating the test facility at Mahendragiri, some 20 km from Nagercoil in Tamil Nadu, was reported in the *1980-81 Annual Report*. So this decision was probably taken some time in 1980. Construction finally began only in late 1982. This decision could have been taken two years earlier, around the time when the site was chosen, says Rajan. As a result, when the indigenous engine was ready, the facilities for testing it were still being built.

The first task of the Vikas team was to show that the engine could be made in India. Before the fabrication of the actual engine was cleared, the liquid propulsion team was permitted to build an engineering model. It was intended only to give an idea of its size and dimensions. Since this task got low priority in the VSSC workshops, the fabrication was largely done by a number of small industries in Trivandrum itself. The engine was built in a year or two. It was later kept in the foyer of the main building at VSSC and became an interesting exhibit for staff as well as visitors.

The Vikas/Viking engine has three principal components: the gas generator, the turbopump and the combustion chamber. It uses Unsymmetrical Di-Methyl Hydrazine (UDMH) as fuel and nitrogen tetroxide (N_2O_4) as oxidizer. These propellants are hypergolic, i.e. they ignite spontaneously when brought into contact with each other. A small quantity of the propellants is burnt in the gas generator and the hot gases are cooled by a spray of water to about 600 degrees Centigrade. These hot gases and steam then pass to the turbopump where they drive the turbine. The rotating turbine turns two pumps, one for each propellant. These pumps feed about 250 kg of propellants every second under high pressure. The propellants are injected into the combustion chamber as a fine spray. Their burning gives off hot gases which are expelled through the nozzle, producing about 74 tonnes of thrust in vacuum.

Making the Vikas engine involved high-precision fabrication. A number of industries, including Hindustan Aeronautics Limited (HAL), the Hyderabad-based Machine Tools Aids & Reconditioning, and Godrej were involved in the fabrication. In order to drill the injectors through which the propellants would be squirted into the combustion chamber under high pressure, one of Hindustan Machine Tools' machines was bought and modified. Only one specialized equipment had to be imported for the fabrication of the Vikas engine.

By the time the engine was ready in 1985, the test complex at Mahendragiri had not yet been completed. It was decided to test this first engine abroad. Although it was the French company SEP which had given the technology, they were reluctant to test the Indian engine.

If there was any fabrication flaw or incorrect assembly, there could be an explosion which would damage their test stand. This test stand was still being used by them to test the Viking engines. In the end, however, SEP was persuaded to test the Vikas engine for ISRO.

The tests in France were a nerve-racking ordeal for the ISRO engineers who went there. They had gone with just one engine in hand. So there was no standby hardware in case of any problems. Moreover, as one of them remarked, 'there were lots of people in VSSC waiting for us to fall flat on our faces'. If the Vikas engine had failed, it would have justified the doubts which had been expressed about using the engine in the PSLV. The gas generator, then the turbopump and finally the engine were tested. 'Every day, all of us would get up early, have a bath and offer prayers before going for the tests. We were so superstitious that we always kept the same seating arrangement. Every second of the test felt like an hour and we would be sweating,' recounts one of the engineers who was there.

But it was worth all the trouble. The full duration test of the engine in December 1985 was a complete success. It put an end to doubts about whether the Vikas engine could be made within the country.

All test facilities at Mahendragiri were ready by October 1987 and further tests of the engine and its components were carried out there. The Vikas engine was subjected to a full-duration firing at Mahendragiri in January 1988. In April that year, the engine completed its qualification trials when it was operated continuously for 3 minutes, 30 seconds longer than it would perform in actual flight.

The contract with SEP gave ISRO technology in the form of technical drawings and other details only for the Viking engine. Based on whatever they had picked up in

France, ISRO's engineers would have to design and build a complete stage using this engine. Although there may be differences in the layout of the stage, principally due to the fact that the diameter of the PSLV's second stage is 0.2 metre larger than the equivalent Ariane stage, the parts which went into the former were almost identical to the ones in the latter. The single propellant tank divided by a bulkhead into two separate compartments to hold the fuel and oxidizer was made by L'Air Liquide of France. The technology for making the tank was later transferred to Hindustan Aeronautics Limited.

The stage is as complex as the engine. A thrust frame is necessary to absorb and pass on the force generated when the engine is firing. There are gas bottles to provide the initial pressurization necessary in the propellant and water tanks. Pipes and valves which open and close on command are required. A command system controls the starting and stopping of the engine. A fill and drain system, as its name suggests, allows propellants to be pumped into the tanks and, if necessary, to drain them as well. The 'pogo' corrector prevents oscillations developing in the fluid flow to the engine. The gimbal control system swivels the entire engine along two axes for pitch and yaw control. The roll control system uses thrusters powered by gas bled from the gas generator.

The stage too was subjected to a number of tests. In the final full-duration test in October 1992, the PSLV second stage in its flight configuration was tested for 149 seconds. Knowledge gained in France about the engine and stage systems had been converted under Indian conditions into flightworthy hardware. It was not a mean achievement.

Even before the stage was ready, ISRO had initiated steps to see if the engine could be made by industry. The

Annual Report for 1988-89 said:

> Encouraged by the performance of various components developed in Indian industries for the Vikas engine, efforts are mounted to develop high-tech vendors who would take up the manufacture of the whole engine or its major sub-systems. The development of a few specialised manufacturers for this major system would not only result in scaling up the level of participation of industry in the Space programme but also ensure a sustaining base for supply of these engines in large numbers as required by the future programmes.

A year later, contracts had been signed with two private companies, Godrej and the Hyderabad-based Machine Tools Aids & Reconditioning (MTAR) for the fabrication and supply of the Vikas engines. Subsequently, a similar contract was signed with a state public sector company, Kerala Hitech Industries Limited (Keltec). The contracts minimized routine production at ISRO facilities, leaving them free to meet development requirements. ISRO has been able to ensure that these companies are able to deliver the engines to its quality standards and schedules. Godrej and MTAR are said to have already supplied twenty-two engines and Keltec four.

At present, though, some part of the assembly of the Vikas engines is done by these companies, the final assembly and testing is done by ISRO itself. Attempts are under way to see if the companies can supply the fully assembled engines.

The principal improvement to the Vikas has come about through indigenization of what is known as the throat insert. Made of silica embedded in a phenolic resin, this material protects the 'throat', the narrow region between combustion chamber and nozzle, from the hot gases whose temperature exceeds 1,000 degrees Centigrade. A Vikas engine with the indigenous throat insert was tested for 200 seconds in July 1995, the longest duration for which Vikas has been tested.

But the development of the indigenous throat insert was carried out by VSSC's Propellant Fuel Complex and composites people. It is not a development for which the liquid propulsion engineers can claim credit. In fact, by many accounts, the Liquid Propulsion Systems Centre was lukewarm about indigenizing the throat insert when it could be imported. It was reportedly only after the item became embargoed under the Missile Technology Control Regime that the indigenous effort was supported.

The Vikas with the new throat insert has been used to upgrade the PSLV's second stage. With a suitably bigger tank and longer stage, the second stage can carry 40 tonnes of propellant. The additional 2.5 tonnes of propellant permits the Vikas to operate for 10 seconds longer, increasing the PSLV's payload capability in polar sun-synchronous orbit by about 50 kg. This version of the PSLV second stage flew in the PSLV-C1 launched in September 1997. The Vikas will also power the four liquid strap-ons of the Geosynchronous Satellite Launch Vehicle.

PSLV's fourth stage

When the PSLV project began, it was decided that a liquid propellant engine would be built for the roll control of the massive launch vehicle during the operation of the first stage. When the advantages of having a liquid fourth stage for the PSLV became apparent, it was found that such a stage could be made by improving and clustering two of the engines being developed for first stage roll control. That

PSLV's fourth stage powered by two pressure-fed liquid engines

way, the development of yet one more engine could be avoided.

Surprisingly, the initiative to have a liquid engine for the first stage roll control, and later to use it for the fourth stage, seems to have come not from the liquid propulsion group, but from the PSLV project. The liquids group was reportedly reluctant to undertake such development. According to the late S. Srinivasan, the first PSLV project director, the liquids group had to be, in fact, pushed into it. The liquids group was unwilling to take up any development other than the indigenous production of the Viking engine and stage, he said in an interview for this book.

The decision to use the roll control engine for the upper stage increased the demands on the engine and greatly complicated its development. Although the fourth stage — or the Liquid Upper Stage (LUS), as it is called — is the smallest of the PSLV stages, it provides a quarter of the velocity needed to put the satellite into orbit. For roll control purposes, the engine needed a combustion efficiency of only 90 per cent. For its use in the fourth stage, however, the PSLV project reportedly demanded a minimum combustion efficiency of 96 per cent. In addition, while the engine needed to operate for less than 3 minutes for roll contol, in the fourth stage it had to be capable of firing continuously for about 7 minutes.

Consequently, developing an engine suitable for use in the fourth stage turned out to be much more difficult than making one for roll control purposes alone. In fact, the problems of developing such an engine appeared at one point to be almost insurmountable.

One major problem which had to be overcome to get the required performance was ensuring effective cooling. The fourth stage engines were to be regeneratively cooled. In regenerative cooling, one of the propellants is passed through channels around the combustion chamber to cool it down. A simple regenerative cooling had been tried out in the prototype developed for APPLE satellite's apogee boost module. But the regenerative cooling for the PSLV's fourth stage engines needed to be much more efficient.

When other countries made regeneratively cooled engines, these channels were cut into the metal wall of the combustion chamber. This apparently could not be done for the PSLV fourth stage engine because suitable equipment was not available. So an ingenious solution was found. A wire was wound helically around the combustion chamber and welded into place. Afterwards, an outer jacket was fixed.

Problems arose in the welding of this outer jacket. Imperfections in the welding resulted in tiny beads forming on the inner surface of the jacket. These beads partially blocked the propellant channels which were only about a millimetre wide. The impaired propellant flow resulted in insufficient cooling. Engine after engine failed after the heat from the hot gases burned through the walls of the combustion chamber.

In desperation, other alternatives were tried. The *1987-88 Annual Report*, for instance, speaks of an ablative cooled engine being designed. But, in the meantime, the reason for the repeated failures with the regeneratively cooled engines had been carefully analysed and traced to defects in welding. The welding techniques were then improved. As an issue of *Propulsion Today*, the LPSC house journal, relates:

> The initial failures of the engine during developmental tests were traced to weld protrusions in the regenerative coolant flow passa ge. Strict quality control of machine-set parameters for EB [electron beam] welding and use of better EB welding facility at GTRE [the Gas Turbine Research Establishment], Bangalore helped in solving this problem.

As *PSLV Progress*, the house journal of the PSLV project, relates in its very first issue, a string of failures were encountered with the fourth stage engine and it took nearly 150 test-firings before arriving at the final configuration which successfully withstood a 60 seconds initial firing followed by a 530 seconds long duration test at Mahendragiri.

No other liquid engine or solid motor that ISRO has ever built required so much testing. Although the development of this engine had begun sometime in 1983, it was only in March 1988 that the engine was successfully tested for 425 seconds, the duration it would operate in flight. But from then on progress was rapid. By 1989, the full stage was being test-fired.

After the PSLV's third stage is separated, the rest of the vehicle coasts along for close to 5 minutes before the fourth stage engines are ignited. Under these conditions of weightlessness, the propellants can float away from the inlets of the pipes. So the tanks are equipped with devices that ensure that propellant is available when the time comes to start the engines. These fourth stage engines are pressure-fed, relying on pressurized helium gas to force the propellants from the tanks and into the combustion chamber. Both engines can be swivelled and, therefore, be used for controlling the vehicle's orientation along all three axes.

Each of the two engines delivers about 750 kg of thrust. Although the fourth stage engines typically operate for about 400 seconds, the stage carries sufficient fuel to keep the engines firing for another 15 to 20 seconds. So the fourth stage is capable of compensating for any underperformance of the lower stages or errors in guidance.

The only problem with this stage occurred during the fourth flight of PSLV in September 1997. On this flight, there was a sharp fall in the pressure of helium gas used for pressure-feeding the propellants to the two engines. The pressure drop occurred immediately after ignition of the fourth stage and lasted for 21 seconds. The result was a shortfall in final velocity of 1.7 per cent from the 7.4 km per second needed to achieve the 817 km circular orbit. The IRS-1D satellite was, therefore, left in an elliptical orbit, 820 km by 300 km.

The problem was traced to pressure oscillations created by the erratic functioning of two pressure regulators which controlled the flow of high-pressure helium to the propellant tanks. The pressure oscillations resulted in the opening of a relief valve and helium leaked out. Analysis and ground tests showed that increasing the volume between the pressure regulators would damp the oscillations and prevent recurrence of the problem. For the next launch in May 1999, 300 cc of buffer volume was introduced between the two pressure regulators.

Inertial guidance

The PSLV would be carrying operational Indian Remote Sensing (IRS) satellites, not experimental payloads as the SLV-3 and the ASLV did. As such, the requirements for its guidance system were necessarily more stringent.

The IRS satellites had to be in polar sun-synchronous orbit. Precise orbital height and inclination is crucial to achieving such an orbit. The PSLV had to inject the satellite into a circular orbit which could not be off by more than 35 km from the specified orbit. The orbit's inclination had to be within 0.1 degree of the specification. Greater the deviation of the satellite's orbit from these specifications, more would be the fuel used up by the satellite's onboard thrusters to correct the orbit. This consumption of fuel would reduce the satellite's life.

In the PSLV, the height, velocity and angle of injection had to be better controlled so that a more accurate orbit was achieved. Unlike in the ASLV, where the closed loop guidance system was jettisoned along with the third stage, the PSLV's closed loop guidance was active right up to and beyond the time of injection.

The ASLV's closed loop guidance had the added disadvantage that a certain level of performance from its fourth stage had to be assumed. There was no way of shutting off the motor when the required velocity was reached. It was for this reason that the Liquid Upper Stage had been substituted for the solid motor originally planned for the PSLV's fourth stage. The fourth stage engines can be stopped when the required velocity is attained. When there was a slight over-performance of the lower stages during the PSLV flight of May 1999, the guidance system was able to compensate and ensure that orbital accuracy

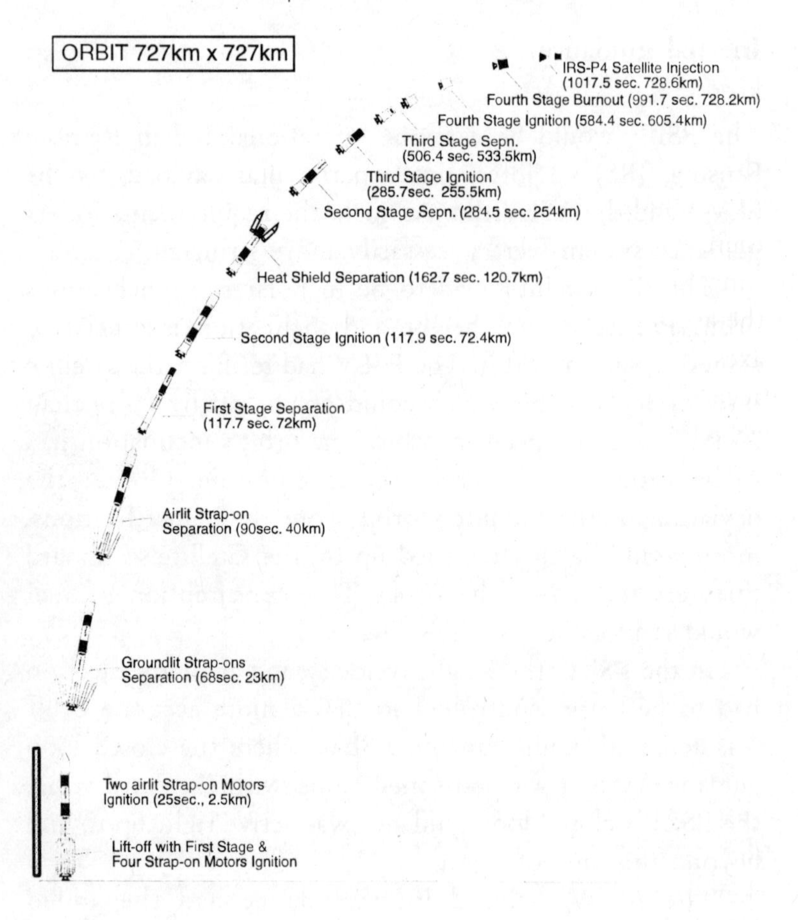

PSLV-C2 Flight Profile

did not suffer as a result. If, on the other hand, the lower stages underperform, the liquid fourth stage carries about 50 to 60 kg of extra propellant so that it can be fired for a slightly extended duration to compensate.

The penalty for PSLV's guided injection was that the weight of its equipment bay and its electronic packages reduced the launcher's payload by an equivalent amount. The ability to use a strapdown system, instead of the stabilized platform used in the SLV-3 and ASLV, helped keep down the weight penalty in this respect.

Stabilized platforms, however, have the advantage that the accelerometers always remain parallel to the three axes of the inertial frame. So deriving the acceleration and thereby the velocity along these three axes is relatively simple. In PSLV's strapdown system, on the other hand, the accelerometers are aligned along the pitch, yaw and roll axes of the launcher. So the accelerometers and the axes of acceleration they sense twist and turn with the launch vehicle. Using powerful microprocessors, however, it is possible to compute the velocity along the three inertial axes from the data provided by the accelerometers and gyros.

The designers of PSLV's guidance system went to quite extraordinary lengths to provide redundancy, detect failures or malfunctions and ensure the continued safe operation of the system throughout the flight. This approach is exemplified by a number of ingenious schemes, starting with the inertial sensors themselves. The three dynamically tuned gyros (DTGs) used in the PSLV can give information on the rotation of about two axes and were arranged in a skewed fashion. With this arrangement, there would be one DTG providing the primary information about each of the three axes. The data from the primary DTG could, however, be cross-checked using the outputs of the other two DTGs. This way, a faulty DTG can be quickly identified. The navigation software is able to detect failure of a gyro and reconfigure the system to use the outputs from the other two gyros instead. As long as only one gyro fails, the mission can safely continue.

There are three accelerometers along the thrust axis, with a majority voting system to identify the failure of any one of them. There is, however, only one accelerometer each for the two lateral axes. In the event of any one of them failing, the acceleration for that axis would be assumed to be zero. In a published paper, B.N. Suresh, currently VSSC's deputy director in charge of Avionics, pointed out that since lateral accelerations were generally small, assuming zero acceleration would lead to higher error in injection but the mission itself would be saved.

The PSLV's guidance system has two chains, with one chain serving as a back-up. Each chain has two processors, one to handle the navigation function and another for the guidance and control functions. In PSLV, the Motorola 68000 processor has been used so far. The arrangement is shown on p. 207.

The navigation processor of one chain is not only linked to its own guidance and control processor but also to that of the other chain. With such 'cross strapping', even if the navigation processor of one chain fails and the guidance and control processor of the second chains also follows suit, the guidance system will still function properly. While redundant chains are common, launch vehicles for unmanned applications generally do not use such cross strapping. One explanation given has been that other launch vehicles these days had a single processor to handle all computations for navigation, guidance and control functions and so did not need cross strapping. With more powerful space quality microprocessors now available, ISRO is also planning to move to a single-processor system.

The price of such cross strapping is greater software complexity to continually synchronize operations of the two chains, detect faults and seamlessly switch from one processor to another. Software complexity increases the difficulties of ground-testing to make sure that all defects

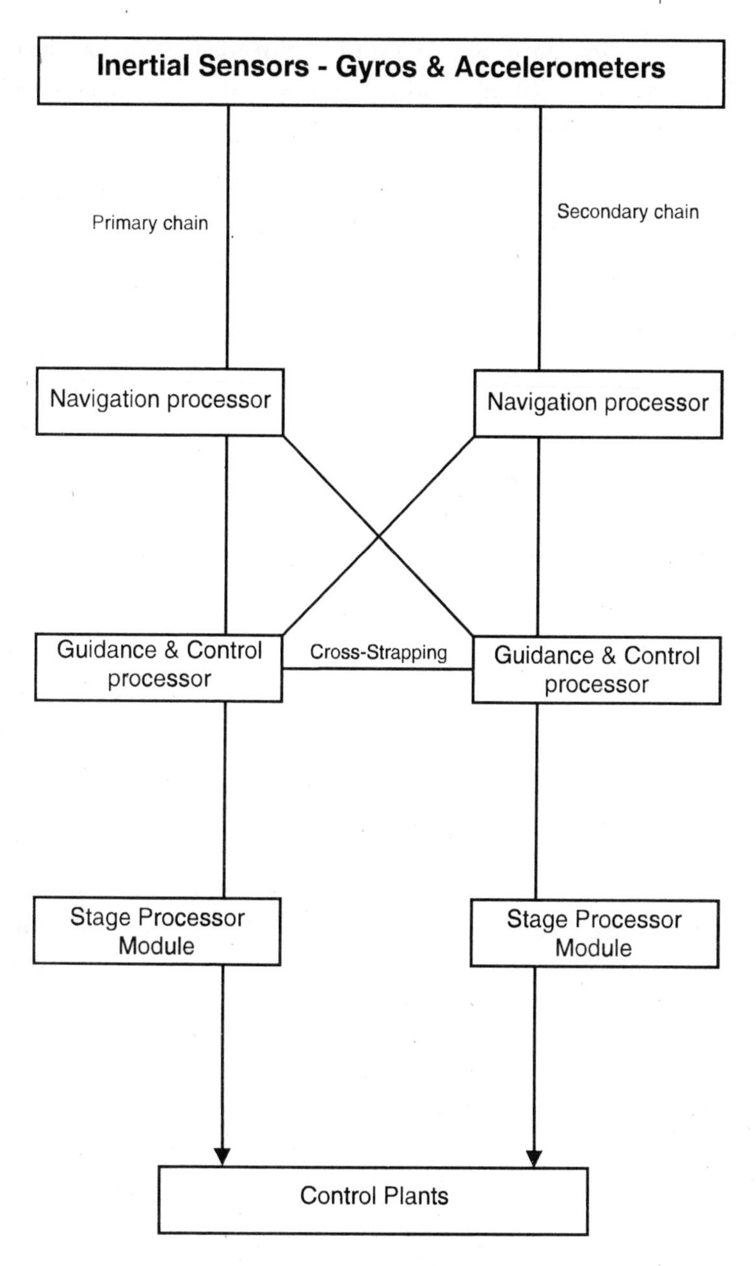

Schematic Diagram of PSLV's Onboard Guidance System

in the software are detected and removed. Getting such a complex system flying, and flying successfully, is therefore no small achievement.

In all the PSLV flights so far, a situation has never arisen when one processor on each of the two chains failed and cross strapping was needed. On one flight, though, one processor was reportedly shut down towards the very end of the mission and the equivalent processor on the redundant chain was then used. But the failure could have been handled with redundant chains without cross strapping.

In the PSLV too, the initial part of the flight is with open loop guidance. The closed loop guidance begins during the operation of the second stage, at a height of about 125 km, and thereafter controls the launch vehicle's trajectory and the injection of the satellite into orbit. After injection, it also changes the orbit of the spent fourth stage to prevent it from colliding with the satellite.

A very similar system will be used in the Geosynchronous Satellite Launch Vehicle (GSLV), the first of which is to be launched in 2000 with a Russian cryogenic stage. The GSLV and the cryogenic stage will be dealt with in a later chapter. In the GSLV, of course, the guidance system will have to follow a very different trajectory so that the Insat communication satellite is injected into the geostationary transfer orbit.

Industry involvement in PSLV

Right from Sarabhai's time, every ISRO chairman has given importance to going out to industry. The PSLV, being an operational vehicle which ISRO intended to continue launching, offered the opportunity for many large contracts. Some forty-six major industries and institutions were involved in the SLV-3 programme. In the PSLV, on the other hand, over 150 industries are involved, public sector and private, large companies as well as small ones.

Apart from the large number of industries participating, it is estimated that something like 70 per cent of the money spent by ISRO on the PSLV currently goes to industry. That would mean that out of the Rs 666 crore currently sanctioned for the PSLV Continuation Project, something like Rs 460 crore would go to industry.

The PSLV involves many fabrication contracts. The Hindustan Aeronautics Limited is one of the major companies involved, supplying several items. Larsen & Toubro and Walchandnagar Industries make the motor cases out of the maraging steel sheets. The production of the entire Vikas engine has been contracted out to Godrej, Machine Tools Aids & Reconditioning (MTAR) and the Kerala Hitech Industries Limited (Keltec). Many other industrial units, big and small, carry out a variety of fabrication tasks.

The PSLV programme also saw contracts being awarded for the production and supply of bulk materials. These were crucial to the launch vehicle programme. Imports would have been impossible and, even if allowed, would have left the programme vulnerable to embargoes. So ISRO made sure that the technology for their production was developed and then ensured their production in industry. The indigenization of maraging steel, HTPB and liquid propellants stand out as examples of ISRO's foresight in this matter.

Maraging steel

Maraging steel is not, strictly speaking, a steel. Apart from iron which is its major ingredient, most steels contain varying quantities of carbon as well as other metals. But maraging steel has only trace amounts of carbon. Instead, its principal ingredients, apart from iron, are nickel, cobalt and molybdenum as well as small amounts of aluminium and titanium. The first step in using this material was to make large metal blocks, termed ingots. The key here is to control the level of impurities which would affect the maraging steel's properties. In addition, despite constraints in the capacity of furnaces available, ingots weighing several tonnes each had to be produced. Only Mishra Dhatu Nigam Limited (Midhani), which had just been set up, had the facilities to make maraging steel on an industrial scale. It is still the sole producer in the country.

These ingots then had to be converted into rings and plates. The plates had to be rolled and then welded to make the cylindrical casing. The rings would be attached to each end of this cylinder, providing both stiffening and the mechanical means to join the segments together. Maraging steel acquires its strength and toughness through a heat treatment process. Before such treatment, it can be quite easily forged, shaped and machined. The most important achievement of the Indian team was coming up with a heat treatment procedure which gave maraging steel the properties specified by ISRO.

The difficulties of doing so is illustrated by the experience in making the maraging steel rings for the first time. Equipment was not available within the country to make rings, the largest of which was nearly 3 metres in diameter and about 3 tonnes in weight. So an international tender was floated. Although Japanese and French companies applied, they were unable to guarantee the fracture toughness which ISRO demanded. The contract went to the well-known German company Krupps, world leaders in ring manufacture. They promised to meet the fracture toughness criterion. Two trial rings they made from Midhani's maraging steel blocks met the specifications. Krupps was then given a production contract and proceeded to make sixteen rings in the first campaign. To their horror, these rings did not have the required fracture toughness.

When Krupps consulted a well-known Swiss metallurgist, they were reportedly told that the fracture toughness for the rings fell in the range which could be expected in manufactured products. In other words, increasing their fracture toughness was difficult. Since the rings represented both a sizeable financial outlay as well as use of a strategic material, the Indians decided to see if the rings could be rescued and a team comprising largely of young people was set up for the purpose. By carefully studying the microscopic structure of maraging steel treated in various ways, they came up with a possible solution in just ten days. Krupps helped, investing considerable money of its own so that the remedial treatments could be tested and proven before work on the actual rings began. In the end, every one of the sixteen rings was successfully retrieved. For making more rings, the team also slightly modified the ring rolling technique and established a heat treatment process which gave both the strength and toughness ISRO demanded. After that, Krupps had no further difficulty supplying the rings to ISRO specifications.

Krupps reportedly stopped supplying the rings a few years back, citing the Missile Technology Control Regime. VSSC and Midhani were, however, together able to help another European company, with no previous experience of maraging steel, make the rings. Facilities have recently been established so that the rings can now be made within the country itself.

The making of plates and welding them posed other problems. Rourkela Steel Plant apparently had serious reservations about taking on the job of converting the maraging steel ingots from Midhani into plates. Rourkela's experience was only with mild steel, not with such high-strength material. In order to meet the specifications for the finished plate, process parameters such as the nature of the atmosphere inside the furnace, temperature at the time of rolling and rolling thickness had to be carefully controlled. By Rourkela's standards, this was a very small job for which they would have to shut down and clean the rolling mill before the maraging steel work could be undertaken. In the end, Rourkela agreed, motivated by the desire to help a national endeavour. The various technical problems were overcome and a process to roll sheets which met the specifications was established. Rourkela has since supplied the maraging steel sheets required by the space programme.

A fresh problem turned up when the sheets were rolled and welded to make the casing. The strength and fracture toughness were inadequate at the weld. A suitable filler wire was developed to be used in the welding. Methods for repair welding and localized heat treatment allowed faulty sections to be reworked so it met the quality requirements. Equipment such as automatic welding machines, ultrasonic testing machines and heat treatment furnace were installed and commissioned at the premises of the two companies which would supply the rolled segments, Larsen & Toubro and Walchandnagar Industries. Both companies have been able to supply the rolled and welded segments to ISRO's specifications.

The Maraging Steel Indigenization Programme was successfully completed by the end of 1983. The *1984-85*

Annual Report of the Department of Space stated: 'maraging steel has been fully indigenised, rolling of plates was established and ring forging techniques have been finalised.' A remarkable multi-institutional team effort had been able not only to develop the technology for everything from the maraging steel itself to the finished products, but also put it into production.

HTPB

Nocil was persuaded to undertake production of HTPB. As with the rolling of maraging steel at the Rourkela Steel Plant, HTPB was a small job for Nocil. It was used to production volumes of tens of thousands of tonnes per annum and ISRO's requirement of HTPB was reportedly just 30 tonnes per annum, points out P. Sudarsan, then head of the Technology Transfer Group and director for Industry Cooperation at ISRO headquarters.

Nocil was well placed to take up production of HTPB. The basic raw materials, butadiene gas and iso-propanol, were being manufactured by the company. Since it was manufacturing raw materials for plastics, such as PVC, it had experience of the industrial-scale processing involved. VSSC's polymer specialists helped Nocil's own R&D team commission a new reactor vessel and use the pilot plant facility to produce HTPB. The contract with Nocil for technology transfer and buy-back of the HTPB was signed in 1984-85. The plant was commissioned in June 1987.

Liquid propellant production

As with HTPB for solid propellant, ISRO was able to get industry to produce the fuels needed for its liquid engines,

Unsymmetrical Di-Methyl Hydrazine (UDMH) for the Vikas engines and Mono-Methyl Hydrazine (MMH) for the PSLV fourth stage engines.

The decision on which route to follow for the synthesis of UDMH became a contentious issue. So much so that, as in the case of maraging steel, a national committee was set up and asked to recommend what ought to be done. There were three different synthesis routes in contention, each associated with a different group. Muthunayagam's group had developed the traditional synthetic pathway involving nitroso dimethyl amine (NDMA). NDMA is so carcinogenic that this route was abandoned, although it was the cheapest method for making UDMH.

Gowariker's group developed a non-carcinogenic way of making UDMH from urea. The *1977-78 Annual Report* speaks of scale-up studies for this process being carried out at the Propellant Fuel Complex. In the meantime, an alternative chloramine route was being attempted in M.R. Kurup's group and was making slow progress. The national committee decided to hedge their bets and recommended supporting both methods.

A contract was signed in late 1980 with Indian Drugs & Pharmaceuticals Ltd (IDPL), Hyderabad, for producing and supplying ISRO with UDMH made through the urea route. The UDMH plant at IDPL was commissioned in 1983-84. But yields remained persistently low and ultimately the contract was terminated. ISRO had invested Rs 60 lakh in new equipment needed for the plant. After the contract was ended, the equipment was taken back.

For the chloramine route, another company, Andhra Sugars was approached. The company established a 12 tonne per year pilot plant and later a 90 tonne plant at Tanuku in Andhra Pradesh. The pilot plant was then used to produce MMH, the fuel used by the PSLV fourth stage engines. The UDMH-MMH plants were formally inaugurated by the Vice-President on 4 July 1988. Since then, Andhra Sugars has supplied these fuels to the space programme.

In the early Eighties, after the indigenous fabrication of the Vikas engine had started, ISRO wanted to import some UDMH. The French and the Japanese are said to have demanded an exorbitant price. However, when Andhra Sugars was preparing to manufacture UDMH, the same French and Japanese companies reportedly tried to dissuade it from doing so. They suggested that Andhra Sugars buy UDMH in bulk from them and then sell it to ISRO. But Andhra Sugars rejected the idea.

Hindustan Organic Chemicals Ltd, a public sector company, received the contract to supply nitrogen tetroxide, the oxidizer used by the Vikas engines. The PSLV fourth stage engines use Mixed Oxides of Nitrogen (MON-3) as oxidizer. MON-3 is made by adding a small quantity of nitric oxide to nitrogen tetroxide and a plant has been set up for this purpose at Mahendragiri.

Ground facilities and assembly

The SLV-3 had been assembled horizontally in a building and then taken to the launch pad. The PSLV would, however, be integrated vertically. The principal reason for this is to eliminate the stress on the vehicle when raising it from the horizontal to the vertical position. Although the Russians assemble their launchers horizontally, their launch vehicles are liquid fuelled. So the propellants, which account for three-quarters of the weight, need to be filled only after the launcher has been erected vertically. The PSLV, on the other hand, would weigh about 250 tonnes, even without liquid propellants.

Having decided to integrate the PSLV vertically, ISRO opted to do the assembly right on the pad. There would be a mobile service tower (MST) providing protection during assembly, with platforms for access to the launcher at various levels, cranes, and a clean room at the very top where the satellite would be integrated with the launcher. The service tower is rolled back from the launcher shortly before launch.

The other option of having a separate vertical assembly building and a transporter to move the fully assembled launcher to the pad would have been much more expensive, according to G. Madhavan Nair, who took over as the PSLV project director from S. Srinivasan. Others point out that the tower could keep the launch vehicle protected from high winds. This was an important advantage since Sriharikota is frequently battered by cyclonic storms. But the second launch pad which ISRO is putting up at Shar will have a vertical assembly building where the launch vehicle will be integrated before being rolled out to the pad.

Building the mobile service tower for the PSLV was no easy task. The MST is a steel structure, 76 metres high and weighing 3,200 tonnes. The MST moves on twin rail tracks, pulled by four bogies, one at each of the tower's corners. Eight hydraulic jacks, two under each bogie, raise the whole MST and hold it anchored in position. Since the east coast was prone to cyclones, the MST has been designed to withstand winds up to 160 kilometres per hour and gusts up to 230 kilometres per hour for 3 seconds when anchored.

The bogies and their wheels were a technology in themselves. Each of the thirty-two wheels, sixteen of which are driven by hydraulic motors, have to withstand a load of 125 tonnes. When the Indian Railways were approached, they reportedly said that they would not be able to build bogies which could take such loads. So an order was placed for these bogies with a US company. The US government, however, blocked the sale to India and stopped even the company's Canadian subsidiary from supplying them. ISRO then held a detailed design review, ostensibly to make sure that the company's design would meet Indian requirements, got all the design details and proceeded to make the bogies in India itself.

Another major development was a high-precision radar, called the Precision Coherent Monopulse C-band radar (or PCMC Radar). It was designed and developed by ISRO with the help of Bharat Electronics. In the early days, NASA and CNES of France had provided radars to track sounding rockets. But, by the time SLV-3 was launched, ISRO, with the help of the Tata Institute of Fundamental Research (TIFR) and the Electronics Corporation of India Limited (ECIL), had developed a C-band tracking radar. This radar was used for the SLV-3 and ASLV launches.

The PSLV needed to be tracked much more accurately because it skirts Sri Lanka. Any departure from the trajectory which might result in the launch vehicle's stages falling on land had to be detected early. The vehicle would then have to be destroyed by activating its self-destruct systems. The PCMC radar was developed to meet this requirement. It can track a launch vehicle up to 3,200 km away. Currently, three PCMC radars have been deployed, two at Sriharikota and one in Mauritius.

Once a launch campaign begins, Sriharikota starts to buzz with activity. There can be some 1,000 people working for the launch. About half of them would have come from other ISRO centres. The assembly of the PSLV begins with the stacking of the segmented first stage. Then the six strap-ons are attached. The liquid second stage is brought by road from Mahendragiri and placed on top of the first stage. The solid third stage, the liquid fourth stage and the equipment bay are integrated into a single module before being mounted above the second stage. Then the satellites are mated to the launcher and finally the heat shield is closed.

A variety of explosive devices have to be installed. These pyrodevices perform functions ranging from separation of spent stages to systems for destroying the launch vehicle if it goes off course. Technology for these has been developed in the course of India's launch vehicle programme. A safe arm has been developed which can prevent the accidental ignition of solid motors as well as of explosive devices (such as those used for stage separation).

Integrating the launch vehicle is a painstaking process. A single wire connected wrongly can be enough to cause failure. For the PSLV, an elaborate checkout system using several computers networked together was established. Various checks and tests are carried out as integration proceeds to make sure that work has been carried out correctly. Once the integration is completed, the checkout system continuously monitors the health and status of the launch vehicle's various systems and sub-systems. It is also used for loading propellants and other pre-launch operations.

The launch campaign for the first PSLV flight began in May 1993, although it was launched only in September that year. Currently, however, the PSLV launch campaign takes about fifty days.

The PSLV gets off the ground

If they were placed side by side, the PSLV would tower

over its predecessors, the SLV-3 and the ASLV. It stands 44 metres high, almost the height of an eleven-storey building, twice the length of the SLV-3 and ASLV. It weighs 292 tonnes at launch, seventeen times more than the SLV-3 and nearly eight times as much as the ASLV. At lift-off, PSLV's giant first stage and four strap-ons (the remaining two strap-ons are lit during flight) together generate 728 tonnes of thrust, six times the power delivered by all four engines of a Boeing 747 at full throttle. While the SLV-3 could put a 40 kg satellite into a 400 km orbit and the ASLV 110 kg, the PSLV can take 3,000 kg into the same orbit.

The Indian Remote Sensing (IRS) satellites have to be put in a polar orbit which circles the earth close to the poles. But from Sriharikota, neither a due north or a due south launch is possible. A due north launch would have the spent stages falling over India. A due south launch would risk having some of the spent stages fall on Sri Lanka. By international law, India is responsible for any loss of life or property caused by its rockets. So the PSLV is launched south-eastwards and then turned south after it has cleared Sri Lanka. If a due south launch had been possible, the PSLV could carry about 200 kg more payload to polar sun-synchronous orbit.

This fact was known from the time that the PSLV was conceptualized. For several years, an alternative Polar Launch Station (PLSN) was seriously considered. When the Srinivasan committee submitted its report on possible PSLV configurations in 1978, it suggested some possible locations for a Polar Launch Station. In the end, however, the idea was given up. The cost and logistical problems of having a separate polar launch station was judged to far outweigh its benefits.

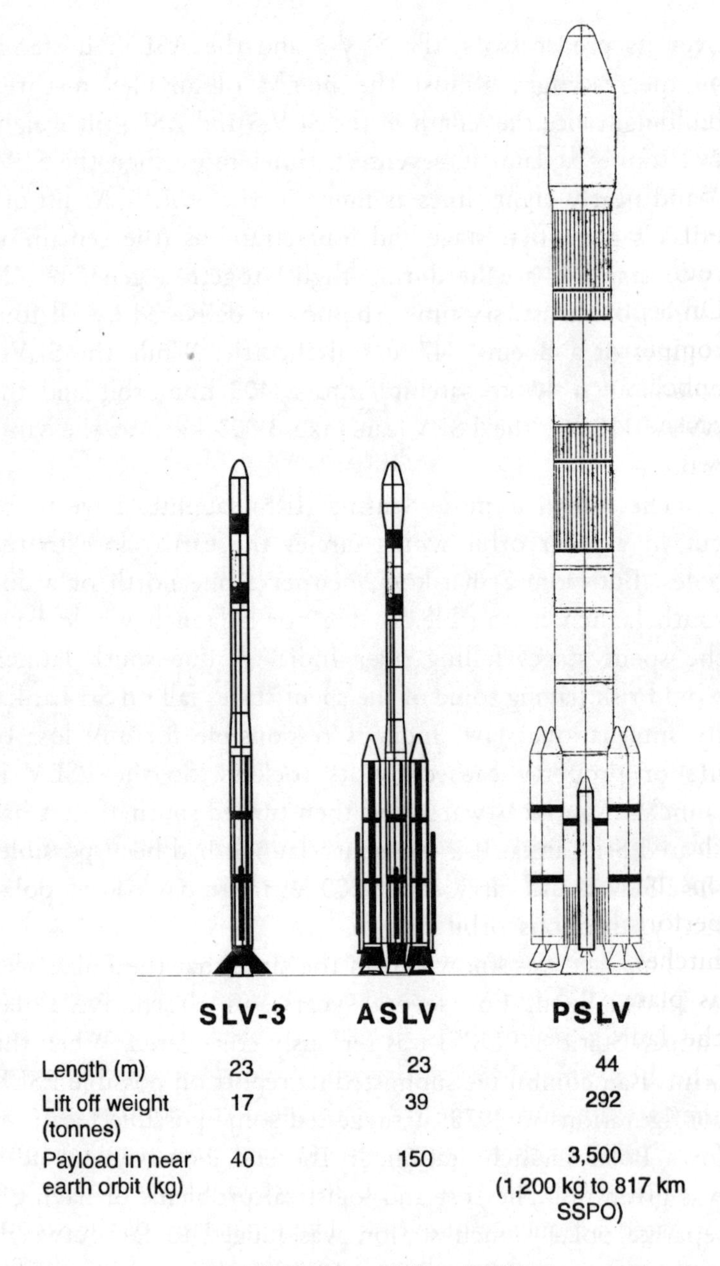

Launch Vehicles : SLV-3 to PSLV

On the morning of 20 September 1993, the PSLV lifted off from Sriharikota for the first time. The PSLV-D1 (the 'D1' indicating that it was the first developmental flight) carried the IRS-1E remote sensing satellite which included the Monocular Electro-Optical Stereo Scanner (MEOSS) built by the German space agency DLR. The final countdown had begun three days earlier, on September 16. On September 19, countdown was halted after the checkout computer detected an anomaly. It turned out, however, that onboard systems were normal and that the anomaly was caused by an error by the checkout computer. The launch was then rescheduled for the next day.

Three seconds before lift-off, the liquid roll control engines of the first stage were ignited. Then, the first stage was fired. In less than a second, its thrust built up and the vehicle lifted off. About a second later, two of the six strap-on motors were ignited and the vehicle rose vertically for 5 seconds.

Inside the Launch Control Centre, a screen showed the course of the launch vehicle superimposed on the planned trajectory. There was no difference between the two and the flight appeared uneventful. The remaining four strap-ons had ignited. The first stage motors and the strap-ons performed well and their separation passed off without any hitches. The second stage, with the Vikas engine, worked as planned. Soon after the second stage began operation, the launch vehicle was turned south on the polar orbit. The heat shield was jettisoned, having served to protect the satellite during the travel through the atmosphere.

Tail-off of thrust as the second stage propellants became exhausted was detected some 261 seconds after launch and initiated a sequence of events. The separation of the second stage was carried out about 3 seconds later and ignition of the third stage commanded 1.2 seconds after that. When the third stage ignited, the vehicle was at an altitude of about 250 km and travelling with a speed of 3.83 km per second. Things went wrong thereafter.

There had been what a subsequent article termed an 'unexpected large disturbance' at the time of the second stage separation. Although the telemetry data radioed down by the launch vehicle showed that the third stage had performed normally and its flex nozzle actuated to the full extent, the disturbance could not be overcome and the launcher went out of control. Nevertheless, the two fourth stage engines ignited as planned and worked steadily for more than 80 seconds, but to no avail. The launcher ended up in the Bay of Bengal.

The national Failure Analysis Committee, which included experts from outside ISRO, submitted its report in January 1994. It said that there were no serious lacunae in the design of the vehicle and the PSLV would have achieved the intended orbit but for a minor deficiency.

The gimballing of the Vikas engines had been halted 3.7 seconds before the ignition of the third stage motor. As a result, disturbances during this period could not be corrected. Two of four retro rockets to push the spent second stage clear of the rest of launcher had failed. This led to contact between the second and third stages, creating the large disturbance which had been noticed.

In spite of these problems, the mission could still have been retrieved but for a small error in the onboard guidance software. The large disturbance at the second and third stage separation had created an 'overflow error' which happens when a computer encounters a value larger than it can handle. Even pocket calculators can face the problem when, for instance, multiplying very large numbers. But while the overflow error is little more than a nuisance in pocket calculators and can be solved by pressing the 'clear' button, it can have unpredictable consequences in a launch vehicle's onboard guidance computer. In PSLV-D1, the overflow error led to the loss of the 'sign bit' which indicates the direction the third stage nozzle has to be swivelled. As a result, the third stage nozzle was flexed in the wrong direction, compounding the disturbance rather than correcting it. The Failure Analysis Committee pointed out that the PSLV design was so rugged that despite the disturbances it had encountered, it would have achieved the intended orbit but for the software error.

Software can be designed to cope reliably when a variable exceeds a preset value and creates an overflow problem. VSSC's control and guidance group was severely criticized for failing to do so. It must, however, be remembered that software routines to protect variables from overflow occupy processor time. All variables used in the navigation and control programme cannot be protected without slowing the programme down unacceptably. So unless there is reason to think that a variable could exceed preset values, it would not be protected.

This lesson was brought home when Europe's powerful new launcher, Ariane-5, was launched for first time in June 1996 and wound up as a spectacular firework display in the skies over French Guiana in South America. An overflow error had led to the main onboard computer and its back-up shutting down.

A joke made the rounds soon after the PSLV-D1 failure. The IRS-1E satellite, which flew on the PSLV-D1, had carried a German Monocular Electro-Optical Stereo Scanner (MEOSS) payload. The SROSS (Stretched Rohini Series) satellite on one of the ill-fated ASLV missions too had carried a German payload. So when the PSLV-D1 failed, it was said that a German payload had sent one more Indian launcher to a watery grave!

The problems of the PSLV-D1, although it ended in failure, were quite minor and easily corrected. A year later, the PSLV-D2 lifted off from Sriharikota on 15 October 1994. After a flawless flight, the Indian Remote Sensing satellite, IRS-P2, was injected into orbit some 17 minutes later. Emotions overflowed in the Launch Control Centre. Tears of joy flowed down the face of K. Kasturirangan, who had taken over from U.R. Rao as chairman of ISRO in March that year. When Kasturirangan and other senior ISRO scientists walked in for the post-launch press conference, they received a standing ovation from the assembled journalists.

The three PSLV launches which followed were successful, showing the essential soundness of the PSLV design. The last develomental flight, PSLV-D3, took place on 21 March 1996. The flight was uneventful and duly did its task of putting the IRS-P3 satellite into orbit.

The Union government had already sanctioned more PSLVs under the Continuation programme. On 29 September 1997, the PSLV-C1 lifted off from Sriharikota, carrying the IRS-1D, identical to the IRS-1C launched in December 1995 by a Russian launcher. A problem with the fourth stage (which has been recounted earlier in this chapter), however, resulted in insufficient velocity to put the satellite into a 817 km polar orbit. The shortfall in velocity left the satellite in an elliptical orbit of 820 km by 300 km. But by using the satellite's own thrusters, the orbit could be corrected.

The then Prime Minister, I.K. Gujral, had come to Sriharikota to witness the launch. Since the shortfall of velocity would have been known immediately, many were critical, even within ISRO, of the organization's failure to own up then and there that a problem had occurred instead of giving the impression that the launch was an unqualified success. It was only in the evening (the launch had been around 10 a.m.) that ISRO announced the shortfall in velocity and the orbit which the satellite was in.

While the criticism is valid, it is instructive to look at the second launch of Ariane-5 in October 1997. Although there was a 1.9 per cent shortfall in velocity, the Ariane-5 launch was proclaimed a success. If the Ariane-5 had been carrying communication satellites, instead of dummy payloads, the additional propellant needed to correct the orbit would have seriously reduced their life. But it was only a couple of days after the launch that the European Space Agency officially admitted in a press release that there had been any problem at all.

When the PSLV-C2 lifted off from Sriharikota on 26 May 1999, it carried not one but, for the first time, three satellites. The principal payload was the IRS-P4 (Oceansat-1), weighing 1,036 kg and with sensors for ocean studies. Riding along with it were two small satellites, South Korea's Kitsat-3 weighing about 107 kg and Germany's DLR-Tubsat weighing just 45 kg. The Prime Minister, Atal Behari Vajpayee, and other dignitaries were present at Sriharikota for the launch. The flight was a complete success.

The PSLV's onboard control and guidance system had been modified so that the IRS satellite was first injected into orbit. The fourth stage and equipment bay (all that would be left of the launcher by that time) would be reoriented by 40 degrees before the Kitsat was ejected. A ball and lock mechanism held the satellite in place during flight. When released, a coiled spring pushed the satellite clear. The procedure was repeated for Tubsat.

A year earlier, in April 1998, ISRO and Arianespace had signed a cooperation agreement for launching micro-satellites weighing up to 100 kg on the PSLV and Ariane-5. If either ISRO or Arianespace secured a contract for a micro-satellite launch but felt unable to carry it on their own launcher, they could transfer the satellite to the other's launcher. Common mechanical and electrical interfaces between the satellite and launch vehicle have been worked out to make such transfer possible, as also common tests for satellite acceptability.

Less than a fortnight after the agreement with Arianespace, ISRO signed a contract to launch the Proba satellite. This 100 kg satellite is being developed by a Belgian company for the European Space Agency and carries an earth-imaging sensor as well as experiments for radiation and space debris monitoring.

Improvements to the PSLV

The PSLV has become what ISRO intended it to be: a workhorse launcher. There are currently six operational IRS satellites in orbit, making it the largest constellation of remote sensing satellites, and four of them were put there by the PSLV. More IRS satellites are planned which too will be launched by the PSLV.

The PSLV is the first ISRO launch vehicle to have its payload capability systematically upgraded. The first launch of the PSLV in 1993, the failed PSLV-D1, carried only an 846 kg satellite, well below its intended payload capability of 1,000 kg. At the second launch in 1994, the PSLV-D2, which was successful, carried just a 804 kg satellite. But by the time of third launch in 1996, the PSLV could carry a 922 kg satellite and ultimately in its fourth launch in 1997, a 1,200 kg satellite. The payload had been increased one and a half times in three years from the PSLV's first successful flight. Many of the improvements made reflect the conservative safety margins adopted by ISRO while designing its launch vehicles.

After the first successful flight of the PSLV, the inert weight of the third stage was reduced by 74 kg and the propellant loading increased. The equipment bay was made lighter by about 25 kg. Steps were also taken to reduce the load on the vehicle when it was travelling through the dense atmosphere. The PSLV-D3 was able to carry a 922 kg satellite, the IRS-P3.

But still more improvements were needed so that the PSLV could launch the 1,200 kg IRS-1D. The propellant loading in the first stage was increased from 129 tonnes to 138 tonnes and the inert weight of the stage reduced. The propellant loading in the liquid second stage was increased from 37.5 tonnes to 40 tonnes. Strap-on ignition sequence was changed from two being lit on the ground and the remaining four in flight to four on the ground and two in flight. The latter had, in fact, been the original plan, but it had been altered after the second ASLV failure to reduce the dynamic pressure on the vehicle. The upper stage weight was reduced by replacing metal with composite materials and miniaturization of the electronic packages in the equipment bay.

Further improvements are planned. A high-performance third stage is being developed. The weight of electronics in the equipment bay is to be further reduced. Some internal structural elements will be made of composites, instead of metal, to minimize their weight.

The PSLV can currently place up to 1,200 kg into 800 km sun-synchronous polar orbit, the launcher's primary mission. The proposed improvements are expected to increase this payload capability to about 2,000 kg. These improvements would also mean that the PSLV would be able to put 1,000 kg into geostationary transfer orbit, instead of the present 850 kg.

Beyond launching IRS satellites

ISRO has at long last, decided to have a separate meteorological satellite in geostationary orbit. Hitherto, ISRO has put the meteorological payload on the Insat communication satellites. Other countries have always had separate communication and meteorological satellites. Metsat, as the new meteorological satellite is called, will be put into geostationary transfer orbit by the PSLV. The first Metsat will probably be launched only in 2002.

Only a few countries make remote sensing satellites and those that do prefer to use their own launchers, unless there is some form of joint development. A Chinese-Brazilian remote sensing satellite was put into orbit by China's Long March vehicle in October 1999. Quite recently, ISRO and the French space agency, CNES, signed a statement of intent for joint development of the Megha Tropiques scientific satellite. The 500 kg satellite will be based on the basic structure of the French Proteus satellite and will be launched by the PSLV in 2005.

With the PSLV, ISRO is now able to offer launches for micro-satellites. But micro-satellites are low-budget ventures and do not yield much by way of revenue. It does, however, give ISRO the experience of dealing with foreign satellite customers and also greater credibility in the launcher market.

The big commercial launch market remains that of putting communication satellites into geostationary transfer orbit. This is a market well beyond the capability of the PSLV as each of these satellites weigh several tonnes.

The emergence of constellations of low earth orbiting (LEO) communication satellites has created a launch opportunity which ISRO would like to exploit with the PSLV. Instead of having heavy satellites in geostationary orbit some 36,000 km away, the LEO communication satellites use a large number of smaller and lighter satellites in orbits much closer to earth. These satellites are used to provide global mobile telephony. The first of these, Iridium, has already begun commercial operations. Iridium uses sixty-six satellites, each weighing about 690 kg and positioned some 780 km above the earth. Constellations of orbiting satellites to provide fast data links for the rapidly growing global internet market are being planned.

The constellations are established by launching several satellites at once on various American, European, Russian and Chinese heavy launchers. Although the constellations usually include in-orbit spare satellites, there may be occasions when a couple of replacement satellites need to be launched quickly in order to maintain communication services. The heavy launchers would be too expensive in such a role and the PSLV would be well positioned to service such a requirement.

In order to make the PSLV more attractive as a launcher for replacement LEO satellites, a smaller fourth stage is being designed which would increase the volume available to carry satellites. The increased volume would make it easier to accommodate multiple satellites on a single PSLV, thereby reducing the launch costs for the satellite operator. Instead of having one large propellant tank divided by a bulkhead into two compartments, one for the fuel and the other for the oxidizer, there would be separate tanks arranged around a single liquid engine. With this configuration, the PSLV would be able to carry up to 3,000 kg into 500 km low earth orbit.

'The continuation programme envisages productionisation and launch of PSLV over the next decade with a capability of touching a launch rate of two to four flights per year, depending on the demand', according to a paper presented at the 1998 International Astronautical Congress by S. Ramakrishnan, the present PSLV project director, and S. Srinivasan. 'While the internal requirement projected is a minimum of one launch per year, it is foreseen that PSLV will make forays into small satellite launch market and can play a role in the emerging LEO constellation maintenance/ replacement launch arena.'

'With the operationalisation of the vehicle and actions for batch production, PSLV is poised to make an entry into the medium lift launch vehicle market,' their paper adds.

But the PSLV may not make the impact which ISRO hopes for in the international launch market, at least in the short term. After Iridium LLC filed for bankruptcy in August 1999, there are serious doubts over whether the Iridium constellation will survive or have to be deorbited. The Iridium failure has also cast a pall of uncertainity over future LEO communication satellite projects.

Penetrating the launch market for small satellites is not going to be easy either. Satellites funded by the US government have to be carried on US launchers. The US company, Orbital Sciences Corporation, has the Pegasus, Taurus and, most recently, the Minotaur launchers in its stable. Similarly, Lockheed Martin has its Athena launch vehicles. Besides, a number of Russian launchers, often converted ballistic missiles, are already in the market offering attractive launch prices. Eurockot, a Europe-based consortium, is marketing the Rockot, a converted SS-19 ICBM. The Dnepr, a converted SS-18 ICBM, has already launched a satellite made by the Surrey Satellite Technology Limited, one of the leading makers of small satellites. A German small satellite was even launched from a Russian nuclear submarine.

In conclusion

The PSLV has an odd configuration by world standards, with alternating solid and liquid stages. Most launch vehicles tend to be powered by liquid engines, with solids for added thrust at lift-off. But however odd its appearance, the PSLV has proved to be a robust design. Despite serious disturbances during the flight, even the first launch would have been successful but for a minor error in the guidance software.

These days successful flights of the PSLV have become routine. It was not so long ago that, after two successive launches of the ASLV ended in the Bay of Bengal, there was a joke that the letters SLV stood for 'Sea Loving Vehicle'. The PSLV has completely turned around the image of Indian launch vehicles, both in the eyes of the nation and of the world.

The PSLV is the realization of Sarabhai's dream of India making its own satellites and lauching them too. ISRO is able to build both satellites and launch vehicles cheaper than developed countries. With the IRS-PSLV combination, ISRO can think of launching remote sensing satellites at regular intervals. The IRS is already the biggest constellation of remote sensing satellites in the world and is set to grow even bigger.

CHAPTER

9

The Saga of the Cryogenic Engine

THE DECISION TO first build the Polar Satellite Launch Vehicle (PSLV) brought with it the question of how ISRO should proceed to acquire the capability to place Insat class of satellites into geostationary orbit. The simplest option turned out to be replacing the top two stages of the PSLV with a cryogenic stage. This vehicle came to be called the Geosynchronous Satellite Launch Vehicle (GSLV).

The critical technology needed for the GSLV is a cryogenic upper stage. A cryogenic engine also runs on liquid propellants. But unlike the Vikas engine which uses earth-storable propellants which remain liquid at room temperatures, the cryogenic engine uses propellants which typically require storage temperatures below –180 degrees Centigrade. All operational cryogenic engines use liquid oxygen (usually abbreviated to LOX) as oxidizer and liquid

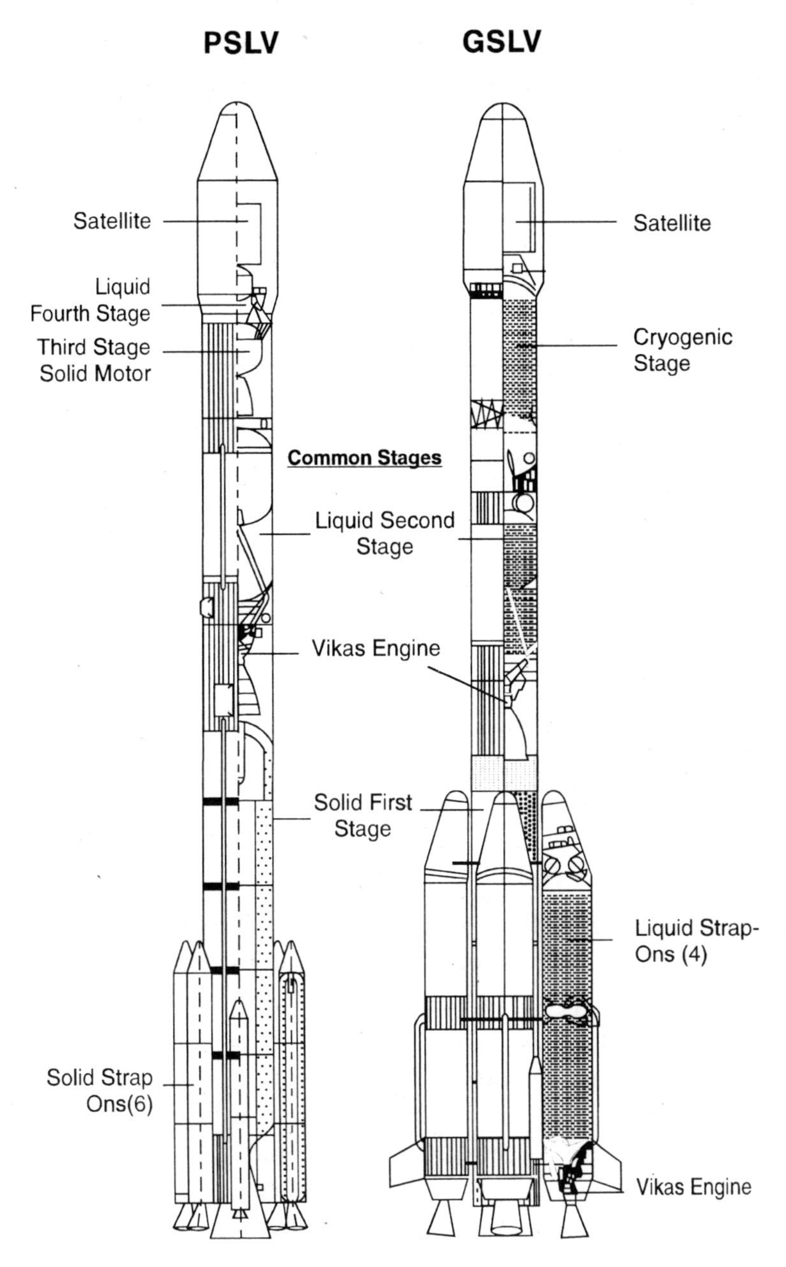
PSLV
GSLV
Satellite
Liquid
Fourth Stage
Third Stage
Solid Motor
Common Stages
Liquid Second
Stage
Vikas Engine
Solid First
Stage
Solid Strap
Ons(6)
Satellite
Cryogenic
Stage
Liquid Strap-
Ons (4)
Vikas Engine

PSLV to GSLV

hydrogen (LH2) for fuel. Oxygen is liquid only at –183 degrees Centigrade and hydrogen at –253 degrees Centigrade. Using propellants which require such low temperatures poses all sorts of problems. These problems complicate the development of cryogenic engines and stages.

The advantage of using cryogenic fuels is their efficiency. To produce 1 tonne of thrust, the Vikas engine would need to burn 3.4 kg of propellant per second. A cryogenic engine can provide the same thrust with only 2 kg of propellant per second. Since the cryogenic stage can be smaller and lighter, a launch vehicle with a cryogenic upper stage is able to carry a bigger satellite. If an upper stage using earth-storable propellants were substituted, more powerful lower stages would be needed to compensate and provide similar payload capability.

Initiatives in semi-cryogenics

Strangely, ISRO's first steps in cryogenics were taken at the initiative of Gowariker, whose primary responsibility was the development of solid propellants. In 1971 he created the Cryogenic Techniques Project and recruited six people for it.

According to P.R. Sadasiva, one of those recruited for the Cryogenic Techniques Project, Gowariker wanted to develop semi-cryogenic engines. Semi-cryogenic engines use liquid oxygen as oxidizer. A suitable grade of kerosene has been widely used as fuel. LOX-kerosene liquid engines have provided the propulsion for many Russian launch vehicles. The world's most powerful liquid engine, the Russian RD-170, uses a LOX-kerosene combination. LOX-kerosene engines have powered several US launchers as well, including the Saturn V which took men to the moon.

LOX-kerosene had two important advantages over combinations such as UDMH-N_2O_4 used in the Vikas. LOX and kerosene are several times cheaper than UDMH and N_2O_4. Moreover, LOX and kerosene are non-toxic while both UDMH and N_2O_4 are highly toxic. The development of semi-cryogenic engines would also have provided a useful intermediate step towards developing a fully cryogenic engine.

In October or November 1971, a project report for developing a semi-cryogenic engine, developing 60 tonnes thrust, was submitted to Sarabhai, says Sadasiva. The project cost was estimated at Rs 3 to 4 crore, a very large sum in those days. Sarabhai suggested that the project be split into two phases. This was done and the revised project report was given to Sarabhai shortly before his death. In fact, just the day before he died, Sarabhai had approved setting up a liquid oxygen plant at a cost of about Rs 16 lakh. But after Sarabhai's death, the project stalled, according to Sadasiva.

The Cryogenic Techniques Project team nevertheless developed a 500 kg LOX-kerosene engine. Like other early liquid engines in ISRO, this one was also quite primitive, using a simple pressure-feeding system. Since there were no control valves to ensure that the two propellants reached the combustion chamber at the same time, a thin foil was used as a burst diaphragm in the two propellant lines. When sufficient pressure built up, the diaphragm would burst and the propellants would gush into the combustion chamber. Unlike the hypergolic propellants which burn on contact, the LOX-kerosene mixture had to be ignited. Lacking a proper igniter, a pyrotechnic device was used instead. But unless the propellants were present in the right proportions, the mixture would not burn properly.

'We used to bring LOX in small containers by jeep from the FACT factory at Alwaye', recalls Sadasiva. FACT used to produce LOX as a by-product from their nitrogen production facility. The containers for transporting the liquid oxgen were not very good and if 100 litres were bought, there would only be some 25 litres left by next morning for the test.

The limitations of the engine, combined with the difficulties of transporting LOX, resulted in the engine being successfully fired only once or twice. These tests were probably carried out during 1973 or 1974. The Cryogenic Techniques Project was transferred from Gowariker to Muthunayagam's Propulsion Division in late 1974. No further work on cryogenics was carried out and the cryogenic team was disbanded.

In 1973, however, it was the Cryogenic Techniques Project which supplied the Vasagam Committee with information about developing a 7.5 tonne thrust fully cryogenic engine and stage. A third stage of this sort was, it will be remembered, included in the launch vehicle configuration recommended by the Vasagam Committee in order to put a 800 kg communication satellite into geostationary orbit.

When options for the PSLV were being considered, the possibility of developing and using a LOX-kerosene engine was examined. In the late Seventies, ISRO set up a study team, headed by M.R. Kurup, to examine the pros and cons of developing a semi-cryogenic engine for use in the PSLV. The acquisition of the Viking engine technology from France made a stage built around the Indian equivalent, the Vikas engine, the strongest contender for PSLV's liquid-fuelled second stage.

The committee recommended developing a 7.5 tonnes thrust LOX-kerosene engine. Clustering four of these engines would give the stage a thrust of 30 tonnes. Although the combined thrust from the four engines was only half that of the Vikas engine, this would be compensated by the LOX-kerosene engines operating longer. The LOX-kerosene stage would carry the same weight of propellant as the Vikas stage.

By building a lower-thrust engine, the difficulties of engine development and, hence, the time needed, would be reduced. The test stand could be simpler, and put up quicker and at much lower cost.

The recommendations of the committee were put out as a comparison between using the Viking engine technology and the new semi-cryogenic engine option, according to H.S. Mukunda, a professor at the Indian Institute of Science who was a member of the committee. There were several factors in favour of the latter option, he says. ISRO needed cryogenic technology and the development of a semi-cryogenic engine would have provided valuable experience. A major programe like the US moon rocket, the Saturn V, had used semi-cryogenic engines because of its low cost. The fact that ISRO's liquid propulsion team which went to France had acquired skills in the design of liquid engines, and not simply fabrication drawings, made the development of the semi-cryogenic engine an economical proposition which could be carried out within the PSLV project schedule. In addition, the committee had recommended that, even if the Viking engine option was chosen, an intense parallel programme on semi-cryogenics should be promoted.

If the Viking technology acquired from France had not been available, ISRO might well have embarked on the development of these semi-cryogenic engines. But ISRO had committed time, manpower and money in securing the Viking technology. The advantages of the LOX-kerosene route had to be weighed against the fact that the Viking technology offered a surer developmental path towards operational capability. There were no overwhelming reasons for ISRO to set aside the investments it had already made in the Viking technology and opt for developing a semi-cryogenic engine. So it is not surprising that ISRO chose to build the PSLV's second stage around the Vikas engine.

But ISRO has paid a heavy price for this choice. No programme for semi-cryogenic and cryogenic technology was initiated. The decision in favour of the Vikas engine created a culture in the liquid propulsion group which favoured import of technology. This import culture in the liquid propulsion group greatly influenced ISRO's fateful decisions when cryogenic technology became necessary for its GSLV.

From PSLV to the GSLV

As the PSLV configuration was being finalized, ISRO began to seriously examine development of cryogenic technology. The cryogenic stage offered the simplest and lowest cost route for building a launcher to put communication satellites into orbit. Simply by replacing the top two stages of the PSLV with a suitably sized cryogenic stage, an Insat communication satellite could be placed in geostationary transfer orbit. In the Eighties, when PSLV development was itself a major challenge, the cryogenic route limited the additional technology which needed to be developed for the GSLV.

Acquisition of technology for a suitably sized cryogenic upper stage was therefore perceived to be crucial quite early.

In December 1982, six months after the PSLV project had been approved, Dhawan set up the Cryogenic Study Team. Its task was to look at alternative configurations for a cryogenic upper stage to replace the PSLV's upper two stages. The study team was also to suggest a strategy for developing a cryogenic engine and stage.

The study team's report, which was titled *Cryogenic System Studies*, was submitted in December 1983. The report ran to fifteen thick volumes, covering every aspect of cryogenic engine and stage development. The report recommended the development of a 10 tonne class of cryogenic engine whose thrust could be upgraded to 12 tonnes.

The *1983-84 Annual Report* is probably referring to this study when it speaks about optimal sizing of a cryogenic engine and stage having been carried out for a future geostationary launch vehicle for the Insat-2 class spacecraft. It is in the same *1983-84 Annual Report* that the GSLV finds mention for the very first time. In a chart displaying ISRO's satellite and launch vehicle missions over the next ten years, the GSLV is shown as being launched in 1991.

Although Dhawan was still chairman of ISRO when the Cryogenic System Studies report was submitted, he seems to have taken the view that decisions on how to proceed with the GSLV and cryogenic engine development ought to be left to his successor. He retired in September 1984 and was succeeded by U.R. Rao who had shaped ISRO's satellite programme.

Since the need for cryogenic technology was clear and, furthermore, a detailed study report had already been submitted, one would have expected ISRO to have promptly initiated measures for developing the technology. The *Cryogenic System Studies* report had recommended the development of a subscale engine producing 1 tonne thrust to acquire the data needed to design the cryogenic engine. This was necessary since such information could not be obtained from published literature. The subscale engine would also help in gaining some experience in fabrication techniques and the handling of liquid oxygen and liquid hydrogen.

However, it was only in 1986 that Rs 16.30 crore was allotted as pre-project funding for such work. Of this, Rs 11.47 crore would be spent on setting up a Cryogenic Engineering Laboratory and building the 1 tonne subscale cryogenic engines. The heat-sink version of the subscale engine was tested during 1988-89 and the water-cooled version in July 1989. Since liquid hydrogen was not yet available, these engines were tested using LOX and gaseous hydrogen (GH2).

The *1986-87 Annual Report* said that a project report for a 12 tonne class cryogenic engine and stage had been prepared. The *Performance Budget* for the following year (1987-88), which would have been submitted to Parliament along with this *Annual Report*, stated that the development of a 12 tonne cryogenic engine would cost Rs 240 crore and was 'under critical evaluation'.

Surprisingly, the *Performance Budget* for the very next year (1988-89) disclosed that 'designs for a cryo engine of 25 tonne thrust (C-25) class are in progress'. But there is no mention of the C-25 engine in the *Performance Budget of 1989-90*, which merely stated that 'configuration studies of an upper stage for GSLV utilising the 12 tonne engine/engines are in progress'. The impression all this conveys is of an organization uncertain about what it wanted to do.

Actually, ISRO was not even sure whether it wanted to develop the technology on its own. It is probable that

ISRO began exploring possibilities for importing the technology around 1986-87. The *Performance Budget for 1987-88* says: 'Experience elsewhere in the world has shown that cryogenic engine development takes about fifteen years. Efforts are being made in ISRO to compress the time schedule to a much lesser period and to realise the cryogenic engine to be available by 1993-94 for the GSLV.'

The Chinese developed their first cryogenic engine, the YF-73, in eight years. The assessment of many of ISRO's own liquid propulsion engineers has been that a cryogenic engine and stage could be indigenously developed in about ten years. But ISRO has repeatedly cited the fifteen-year time-frame for developing the cryogenic technology as its reason for wanting to import the technology, instead of developing it indigenously. The only way to 'compress' the development time to six years (as hinted at in the *1987-88 Performance Budget*) would be to secure the technology from another country.

ISRO tried to get the cryogenic technology from three countries: Japan, the United States and France. Japan reportedly refused point-blank. The negotiations with the United States are said to have been over the RL-10 cryogenic engine used in the Centaur upper stage. The RL-10, developed by Pratt & Whitney, was the world's first LOX-LH2 cryogenic engine and General Dynamics clustered two of these engines in their Centaur stage. But the negotiations with the Americans went nowhere.

ISRO had good reason to look to France for cryogenic technology. It had licensed the Centaure sounding rocket to India, helped in establishing the solid motor test facilities, and provided the technology for the Viking engine. Moreover, according to some of ISRO's senior people in liquid propulsion, France had offered the technology for their HM7 cryogenic engine used in the Ariane launcher for just Rs 1 crore around the time when the Indo-French Viking contract was ending. But by the late Eighties, the French were a lot less accommodating. They are said to have demanded around Rs 1,000 crore for the HM7 technology. Since Ariane was being pushed as a commercial launcher, the French did not want to create a competitor. They demanded restrictions on ISRO using the HM7 technology for commercial launches. According to unconfirmed reports, ISRO was nevertheless inclined to take the technology from the French but was prevented from doing so by the Finance Ministry, which baulked at the price tag.

By 1989, it was clear that ISRO would not get cryogenic technology from Japan, the United States or France. There were only two options left. One was to develop the technology indigenously. The other was to see if the Soviet Union was willing to help out. Both avenues appear to have been pursued at this time. The *Performance Budget for 1989-90* stated: 'The configuration studies for the realisation of indigenous geo-synchronous launch capability for launching Insat-II class satellites by 1993-94 are in the final stage and a project team is being constituted to prepare the Project Report for GSLV.'

The *Project Report for the Cryogenic Engine and Stage* was submitted in 1990. An indigenous cryogenic engine developing 12 tonnes of thrust would be developed and made into a stage carrying 14 tonnes of propellant. The development time was estimated at six to six and a half years. This engine and stage came to be designated the C-12. In November 1990, the GSLV project, which included the development of the cryogenic engine and stage, was approved. The sanctioned project cost was Rs 756 crore and the GSLV was to have its first flight during 1995-96.

Replacing the PSLV's upper two stages with the C-12 and its solid strap-ons with four liquid strap-ons would produce a GSLV capable of putting 2,500 kg satellites into geostationary transfer orbit (GTO).

The Soviet cryogenic engine deal

In the meantime, the discussions with the Soviet Union were going well. In January 1991, ISRO signed a contract with Glavkosmos. For Rs 235 crore paid in Indian rupees, the Soviets would, in a period of six years, provide two flightworthy cryogenic stages as well as the technology to make the stages in India. The contract also included an option to buy three additional stages.

The Soviet engine would produce a thrust of 7.5 tonnes and the stage would hold 12.5 tonnes of propellant. In addition to the main engine, the stage would have two small vernier engines which could be swivelled for attitude control. The Soviet stage would take the place of the C-12 stage and later itself be substituted by an identical stage made in India with Soviet technology. This GSLV, like the one with the C-12 stage, would be able to 'put into orbit a satellite of 2,500 kg' (*Performance Budget for 1992-93*).

But there was a storm brewing which ISRO had conveniently ignored. While ISRO was busy shopping around for cryogenic technology, the Missile Technology Control Regime (MTCR) had come into being in April 1987. It was neither an international agreement nor a treaty but began as a voluntary arrangement between the United States and its six major trading partners. Another twenty-two countries have since become adherents to the MTCR.

The MTCR was intended to prevent the export of technology which could contribute to the development of long-range missiles. The MTCR provides regularly updated guidelines which would be implemented by the adhering nations through their export control laws. Since most of the technology used in launch vehicles is directly applicable to missiles, the MTCR's provisions also affect technology transfer for launch vehicles, such as the cryogenic engine technology. This aspect will be covered in greater detail later.

In November 1990, the very month that the GSLV project was approved, the United States amended its laws so that it could apply sanctions against entities which violated the MTCR. This transformed controls which could be used at the discretion of the US President into a mandatory requirement under US law to apply sanctions against countries and entities violating the MTCR.

Neither the MTCR nor the change in US laws for its enforcement would have affected India's contract with Glavkosmos if the Soviet Union had been as strong as it was during the heyday of Indo-Soviet cooperation. But Mikhail Gorbachev had introduced glasnost (openness) and perestroika (restructuring) in the Soviet Union from the late Eighties. By 1989, conflicts between the central authority in Moscow and the republics had become increasingly rancorous, exacerbated by long-suppressed ethnic conflicts. By 1991, when ISRO signed the cryogenic deal, republics were breaking away from the Soviet Union. The abortive coup of August 1991, which brought Boris Yeltsin to power, led to the Communist Party itself being abolished. By December 1991, the Soviet Union had ceased to exist: the three Baltic states had achieved complete independence while Russia, Ukraine and Belarus signed an agreement establishing the Commonwealth of Independent States.

The only person who seems to have recognized the dangers of depending on the Soviet Union for the cryogenic technology was S. Chandrashekar, then with the Launch Vehicle Programme Office at ISRO headquarters in Bangalore and currently on the faculty of the Indian Institute of Management, Bangalore. 'The cryogenic engine deal clearly violated the provisions of the MTCR and so sanctions required by US law were inevitable. I was convinced that the Soviet Union, in the political and economic situation it found itself, was not going to be able to withstand US pressure. But no one in ISRO was willing to listen to me,' he says.

His misgivings were justified. In May 1992, the United States imposed sanctions against ISRO and Glavkosmos, prohibiting exports to and imports from the two companies for two years. The United States brought pressure to bear on Russia, which had inherited the contractual obligations for the cryogenic deal. If Russia terminated its contract with India, it would be rewarded with access to the commercial launch market as well as participation in the International Space Station. The choice could not have been starker for the Russians. If Russia opted to go ahead with the Indian contract worth only about $120 million, it would be jeopardizing opportunities for their well-developed space industry to earn many times more in hard currency.

The contract staggered on for another year. The *coup de grace* came in July 1993. In a statement to Parliament a month later, the Indian Prime Minister said:

> The Chief of the Directorate of International Scientific and Technical Cooperation of the Russian Ministry of Foreign Affairs handed over a nonpaper to the Indian Ambassador in Moscow on 16 July in which it has been stated that in the context of unforeseen circumstances,

> Glavkosmos finds itself in a situation of not being able to fulfil further its obligations regarding the transfer of technology and production equipment under the Agreement of January 1991. The paper given to the Indian Ambassador invokes the force majeure clause of the January 1991 Agreement as the basis of Glavkosmos resiling from its contractual obligations.

However, official ISRO documents, such as the *1993-94 Annual Report*, give October 1993 as the time when the Russians suspended the contract.

Did the cryo deal breach MTCR?

The blunt answer is that it quite clearly did.

MTCR's stated purpose has been to limit the spread of delivery systems, other than manned aircraft, which are capable of carrying weapons of mass destruction, including nuclear weapons. The delivery systems of concern are specified as those capable of carrying at least a 500 kg payload to a distance of 300 km or more and includes missiles as well as unmanned aircraft. The MTCR covers not just the transfer of full systems, but also of equipment and technology which could be used for making such systems.

Though the MTCR states that it was not intended to impede national space programmes, it is nevertheless based on the premise that technology for launch vehicles can be used for missiles as well. So its implementation in actual practice means that even technology transfers intended for a launch vehicle will not be permitted.

For good technical reasons, no country has ever fielded a missile with a cryogenic engine. Filling and preparing any cryogenic stage for launch is a time-consuming process. Such a stage would make a missile cumbersome to operate and leave it vulnerable to attack. Despite this, the fact remains that India's cryogenic contract fell foul of the MTCR requirements.

The MTCR's Equipment and Technology Annex lists solid and liquid propellant rocket engines having a total impulse of 1.1 mega Newton-seconds under its Category I items. Category I items are those of the greatest sensitivity whose transfer will be permitted only rarely. The total impulse of the Russian cryogenic engine and stage is fifty times greater than the cut-off limit specified in the MTCR.

ISRO should have known this, as also the fact that amendments in November 1990 to the US Export Administration Act and the Arms Export Control Act made it mandatory for the US government to impose sanctions for violations of the MTCR.

Right from Sarabhai's time, ISRO has always been acutely conscious of Western, and particularly American, suspicions of its launch vehicle programme. It vigorously promoted indigenization to protect the launch vehicle programme from becoming vulnerable to embargoes. ISRO's failure to read the writing on the wall with respect to the MTCR and its consequences for importing cryogenic technology therefore stands out in startling contrast.

What happened to the Russian deal?

The Russians managed to have their cake and eat it as well. They renegotiated the contract to their advantage. By not giving the cryogenic technology to India, they were also able to avoid offending the Americans.

Some senior ISRO officials have confided in private conversation that ISRO repeatedly came under pressure from New Delhi not to take too hard a line when the contract was being renegotiated. Russia, it seems, used its leverage as India's largest and most important arms supplier to ensure that the contract continued, even though it was they who breached it.

The contract was first revised to compensate the Indians for the technology transfer which would now not take place. The *1994-95 Annual Report* recorded: 'Within the original cost of the agreement [that is, Rs 235 crore], in addition to the originally planned two cryogenic stages, two flightworthy stages and two ground models will be supplied by Glavkosmos in lieu of the technology transfer. Option has also been exercised to buy three additional cryogenic stages as per the original contract.' The three additional stages would cost $3 million each.

A year or so later, the contract was revised yet again. India's trade with the Soviet Union had been governed by the rupee-rouble trading system. Under this system, Indians could pay in rupees for goods they bought from the Soviet Union and this money would be held in a rupee account in India. The Soviets could then use this money to import goods from India. When the cryo contract was signed in January 1991, India was allowed to make the payment in rupees under this system. In addition, the price of Rs 235 crore in the original contract for the stages and cryogenic technology was definitely a 'friendship price' given by the Soviet Union, considering that France had asked for four times as much and in hard currency.

But after the break-up of the Soviet Union and Russia's growing economic problems, the rupee-rouble trading system was given up. Russia could no longer afford the financial costs of friendship and demanded to be paid in US dollars. The outcome of these negotiations was that the

Rs 235 crore was converted into dollars at the January 1991 exchange rate. The contract was now fixed at $128 million.

ISRO had reportedly paid about Rs 70 crore by May 1992 when the Americans placed their embargo and about Rs 200 crore by 1995 when the negotiations to denominate the contract in dollars were conducted. These rupee payments would be converted to dollars using only later exchange rates. Since the rupee continually depreciates against the dollar, it meant a sharp hike in the rupee price which ISRO would have to pay. When R. Ramachandran, then science editor for the *Economic Times,* and I made an assessment, our view at the time was that ISRO would end up paying about Rs 200 crore more as a result, a doubling of the price in rupee terms. But ISRO officials say that the increase would only be of the order of 30 per cent.

The Russian cryogenic engine and stage

Glavkosmos, which signed the deal with ISRO, was just the marketing intermediary. According to information on the Federation of American Scientists' (FAS) website, Glavkosmos was originally intended as a national space agency to coordinate all civil applications programmes. But other bodies and agencies refused to give up their responsibilities, and Glavkosmos became confined to foreign relations. However, it soon lost even that privilege and, after the break-up of the Soviet Union, became just an intermediary in the space services market.

The Russian cryogenic stage is based on the KVD-1M engine. Mark Wade's online web site, the Encyclopedia Astronautica, lists the other designations of this engine as the 11D56M and the RD-56M. According to the Encyclopedia Astronautica and other sources, the cryogenic stage being supplied to India carries the Russian designation 12KRB.

The KVD-1 was derived from an engine known by the Soviets/Russians as the 11D56, according to Asif Siddiqi, a US-based expert on Soviet and Russian space programme. The 11D56 was originally developed by KB Khimmash for the Soviet Moon programme of the Sixties. At the time, the Soviets were developing a number of major systems as part of the massive N1-L3 project to send man to the moon before the Americans. The 11D56, planned for use in an upper stage of one of the versions of the giant N1 rocket, was the very first Soviet rocket engine to use the liquid hydrogen-liquid oxygen propellant combination.

The 11D56 was ground-tested for the first time in June 1967. The KB Khimmash company also developed a modified version of the 11D56, named the 11D56M (also called the KVD-1M). Development of both engines was terminated in 1977 after the Soviet moon programme was cancelled. Neither engine was ever flown in space. There were four launches of the N1 in 1969-72, but none of them carried the KVD-1 engine, says Siddiqi

The stage being supplied to India carries the Russian designation 12KRB. 'KRB' is the acronym in Russian for 'Space Rocket Block', according to Siddiqi. Russians often use the word 'block' to refer to rocket stages and the '12' in the name may refer to the weight of propellant which the stage carries. While the KVD-1M engines are from KB Khimmash, the stage has been developed by the GKNPTs Khrunichev. According to ISRO, the Russians have extensively ground-tested the engines and stage before being supplying them to India.

Making the 12KRB stage for India appears to have led Khrunichev to develop a similar stage for its Proton rocket.

The Proton Mission Planner's Guide states: 'In the course of a vehicle-development contract with another entity, Khrunichev has developed a liquid oxygen, liquid hydrogen upper stage. This upper stage can be adapted to fly on the Proton booster and could significantly enhance launch-system performance to high-energy transfer orbits.' According to the Mission Planner's Guide, the new cryogenic stage would increase the Proton's payload capability in GTO as well as in geostationary orbit by almost 50 per cent. The KVRB stage for the Proton has not yet become operational. The stage is also expected to be used in the Angara rocket which Khrunichev is developing.

GSLV with the Russian cryo stage

The first of the Russian flightworthy cryogenic stages was delivered in September 1998. The first GSLV with this stage is currently scheduled to fly in early 2001. But instead of putting a 2,500 kg class satellite into GTO, this GSLV is likely to carry only a 1,600 kg satellite. This amounts to an unprecedented decline of 36 per cent in the payload capability. By comparison, although the entirely indigenous PSLV was intended to put 1,000 kg into orbit, it carried a satellite of 846 kg on its first flight, a shortfall of just 15 per cent from its targeted capability.

More important, the GSLV with the Russian stage will not be the big jump in capability over the PSLV which it ought to have been. The improvements currently being implemented on the PSLV will give it a payload capability of about 1,000 kg in GTO. If its six solid strap-ons were replaced with the four liquid strap-ons developed for the GSLV, its payload capability in GTO would be around 1,400 kg.

The GSLV with the Russian stage will provide a payload capability of only 1,600 kg in GTO at the first launch. It is hoped, however, that in later flights this can be raised to 1,800 kg by measures such as lightening the equipment bay and using a composite (instead of metal) payload adapter. This is still well short of the 2,500 kg in GTO which it was supposed to provide. ISRO has ended up paying a high price for an increase of just 200-400 kg over the payload capability it could have got out of a PSLV with four liquid strap-ons.

Two factors are said to have contributed to this decline in the GSLV's payload capability. One has been an increase in the weight of the Russian stage. It has been difficult to get information on the increase in stage weight. The Russians are unquestioned masters of liquid propulsion technology. So the increase in the weight of the stage supplied by them has led to conspiracy theories about the increase being deliberate and having been done to please the Americans. But G. Madhavan Nair, who took over from Muthunayagam in 1995 as director of the Liquid Propulsion Systems Centre (LPSC) and is currently the director of VSSC, dismisses such notions. The Russian approach has always been to have much greater safety margins in their designs and such margins make their stage heavier, he points out.

The performance and weight of the uppermost stage in any launch vehicle is critical in determining its payload capability. As was explained in the chapter on the PSLV, when it was decided that the liquid motor being developed for roll control of the first stage should be clustered and used in the fourth stage as well, the PSLV project team demanded improvement of the engine's performance. After facing countless failures, the required performance was finally achieved. The PSLV project had demanded such performance not out of bloody-mindedness but because otherwise the launcher's payload capability would have been severely curtailed. How is it, then, that the Russians have got away with supplying a heavier stage? Given the criticality of the cryogenic stage's performance and weight, the question needs to be asked as to whether adequate safeguards had been incorporated in the contract to make sure that the cryogenic engine and stage to be supplied met ISRO's specifications.

'Range safety' is also said to have contributed to the decline of the GSLV payload. Range safety has to do with ensuring that spent stages and other debris do not cause loss of life or any damage. As a signatory to the international Liability Convention, India is liable for any damage caused by its space effort in another country.

The spent stages of the SLV-3 and the ASLV, which were launched eastwards, all fell into the Bay of Bengal. In the case of the PSLV, it was only necessary to ensure that the launch vehicle skirted Sri Lanka. But a geostationary launch from Shar has to take into account the Malay peninsula, the islands of Indonesia as well as numerous other islands, including the disputed Spratly Islands in the South China Sea and those in the Pacific. This region also has busy shipping lanes and offshore oilfields. While the first stage and the four liquid strap-ons can be separated safely over the Bay of Bengal, it is the jettisoning of the liquid second stage which is of concern in the GSLV.

Elaborate trajectory analysis is part of the process of configuring a launch vehicle. The range safety constraints involved in a geostationary launch from Shar are well known. How is it that range safety has become a factor in bringing down the payload of the GSLV?

According to one senior person in the launch vehicle programme, the original intention was to drop the spent second stage just east of the Malay peninsula (101-102 degrees East longitude). But there had been an exponential increase in oil-wells in this region since 1991 when the contract was signed with the Soviet Union and it was now too dangerous to jettison the second stage in this region, he says. So the plan is to dump the spent second stage before crossing the Malay peninsula. The need to drop the second stage early has been responsible for much of the payload loss suffered by the GSLV, according to him.

But this explanation seems less than wholly satisfactory. When deciding the GSLV's trajectory, ISRO could have observed stipulations in the United Nations Convention on the Law of the Sea which was opened for signature in 1982 and came into force in 1994. The Law of the Sea gives coastal states sovereign rights in a 200-nautical-mile (370 km) exclusive economic zone (EEZ). The offshore oil wells are situated within this distance. It is precisely for this reason that sovereignty over the Spratly Islands in the South China Sea has become such a contentious issue.

Having a launch vehicle stage fall in the EEZ of another country is inherently risky. The Law of the Sea gives states the right to exploit resources within their EEZ and all major offshore activity, including oil exploration, is concentrated in this zone. A sudden growth of economic activity, which is what ISRO says happened in the case of the GSLV, can therefore occur in an EEZ. Since a launch vehicle can take about ten years to develop and have an active life of five to ten years, it would be necessary to predict events in a EEZ ten to twenty years in advance before thinking of breaching it. So the most fail-safe solution is to avoid having spent stages fall in another country's EEZ.

In fact, in the case of the PSLV, ISRO is said to have decided that none of the launcher's stages should fall within the 200-nautical-mile zone while skirting Sri Lanka. When it came to the GSLV, however, the original projection that it would deliver 2,500 kg in GTO appears to have been made by relaxing this norm and having the spent second stage fall inside the EEZ of another country. If the EEZ had been respected, the GSLV's payload projections would have come down and been more in line with what the launch vehicle is currently able to deliver. Alternatively, if ISRO wanted to have a higher payload capability, it would had to look at a different GSLV configuration.

The overall impression one gets is of an organization in a hurry to clinch the cryogenic deal with the Soviets, not of one interested in examining the gift-horse carefully enough.

No way out except indigenous development

Once the Russians backed out, it became painfully obvious that there was no escaping indigenous development. The initial ISRO reaction was to maintain that much of the cryogenic technology had already been acquired and so an indigenous cryogenic stage could be built reasonably quickly.

In an interview published in *The Hindu* of 26 July 1993, U.R. Rao, then ISRO chairman, stated:

> We have the confidence to build [the] cryogenic engine. The engine will be ready by 1997. One has to redraw the time schedule. The earlier schedule to launch the GSLV was by 1995. And any falling through of the contract with Russia will at most mean a delay of one to

> one and a half years if we use our own engine. We may be even able to speed up.

A year later, Rao reiterated this view in an interview published in October '93-March '94 issue of *Space India*, an ISRO publication:

> We had very tough negotiations with the Russians and we are getting a few more engines from them in lieu of the technology transfers and on the basis of the technology transfer till October 1993, we should be able to go ahead and probably take 2 more years, till 1998, to make it on our own. In the meanwhile the stages procured from Russia would enable us to move fast (in the GSLV programme).

When R. Ramachandran and I spoke to ISRO's engineers, we got a very different picture. The assessment of these specialists was that serious cryogenic development was just getting underway in ISRO. A flightworthy indigenous cryogenic stage was about ten years away. These assessments have proved correct. The piece Ramachandran wrote appeared in the *Economic Times* of 27 July 1993 and mine in *The Hindu* of 4 August 1993.

Even in 1995, a year after K. Kasturirangan had taken over from Rao as chairman of ISRO, the organization's official views do not seem to have changed much. Newspapers of 25 May 1995 carried a newsitem by the Press Trust of India (PTI) of the reply given by the minister of state in the Prime Minister's Office, Bhuvnesh Chaturvedi, to a question in Parliament. According to the PTI report, Chaturvedi said that while there may be 'marginal delays' in the Indian space programme because of the Russian decision, it is not expected to cause any major setback as steps had been taken to indigenize several critical technologies.

The indigenous cryo stage

As mentioned earlier, the official ISRO line had been that substantial transfer of technology from the Russians had taken place by October 1993 when the contract was suspended. ISRO would therefore be able to complete the indigenous stage with only limited delay. So it is not surprising that when the Cryogenic Upper Stage (CUS) project was approved in April 1994 with a budget of Rs 336 crore, its aim was to build a cryo stage similar to the imported Russian one.

Another factor in favour of copying the Russian stage would have been that if serious problems cropped up during development, help might be got from the Russians, albeit, probably, at a price. Special materials required for the cryogenic engine and stage could also be got from Russia (again, at a price). The *1994-95 Annual Report* stated that materials for the engine were being procured and that 60 per cent had already been received. The subsequent *Annual Report* said that procurement of materials had been completed. All the materials needed are now said to have been indigenized.

In a talk in January 1994, A.E. Muthunayagam, the then director of the Liquid Propulsion Systems Centre (LPSC), explained the thinking behind the deal with Glavkosmos. Given the complexity of cryogenic technology, it took even countries with advanced resources more than ten years to develop. The decision to buy a few cryogenic upper stages from Glavkosmos therefore made sense, specially as these were compatible with the ISRO boosters and met the mission requirements. 'This approach has the advantage of shorter development time and lesser risks for an early optionalisation of GSLV. However, for the continued self-reliant operational flights of GSLV, ISRO plans to indigenously develop the cryo upper stages (CUS) with the same specification and interface,' he said.

G. Madhavan Nair argues that experience with the staged combustion cycle used in the Russian stage would be beneficial. In staged combustion, the fuel (in this case, liquid hydrogen) is burnt with a little oxidizer (liquid oxygen) in a precombustion chamber. The hot gases produced drive the turbopumps and are then injected into the combustion chamber, along with more oxidizer. Staged combustion provides the most energy-efficient engine.

As pointed out earlier, ISRO's engineers had, in private, predicted in 1993 itself that an indigenous cryo stage would be ready only around 2003. The official estimate has slowly caught up. The confident assertion that the indigenous stage would be ready by 1997 was soon revised to 1998, then 1999, 2000 and, most recently, 2002.

The GSLV with the imported Russian stages is not expected to give more than 1,800 kg in GTO. Since Russians use considerable safety margins in the construction of their hardware, G. Madhavan Nair believes that some of these margins can be reduced to make the CUS lighter and therefore let the GSLV put over 2,000 kg in GTO. Since the CUS is ISRO's first cryogenic stage, this could prove optimistic. It seems more likely that the CUS performance would be closer to that of the Russian stage, giving between 1,800-2,000 kg in GTO.

The GSLVs equipped with the imported Russian stages have been designated as the GSLV Mark-I. The first launch of the Mark-I is currently scheduled for early 2001. The Russian stage will, however, incorporate Indian control electronics. The GSLV Mark-II will use the indigenous CUS stage and is at present expected to be launched in 2002.

Other indigenous cryo work

After the deal with the Soviets was signed in January 1991, indigenous efforts to develop cryogenic technology seem to have been put on the back burner.

The heat-sink and then the water-cooled version of the 1 tonne subscale engine had been tested in 1989, before the Russian deal. There was also to be a regeneratively cooled version where one of the propellants would be passed around the engine to prevent it from overheating. A 1989 issue of *Propulsion Today*, the LPSC house magazine, stated that the regeneratively cooled version would soon be tested for 500 seconds.

But that does not seem to have happened for another six or seven years. The *1995-96 Annual Report* spoke of 'characterisation of a one tonne class sub-scale cryo engine with LOX and LH2' being in progress and a 1996 issue of *Propulsion Today* noted that the regeneratively cooled version of the subscale engine with LOX-LH2 had been tested for 200 seconds in December 1996. The *1996-97 Annual Report* declared: 'Successful completion of the one-tonne cryogenic engine development programme during the year has been given further fillip towards developing indigenously the cryogenic upper stage.'

The purpose of the subscale engine, according to the 1996 *Propulsion Today* article, was to validate the design parameters and to gain experience in the fabricating and testing of cryogenic engines. After the work done by 1989, why were seven years needed to complete the subscale engine development? The most probable explanation is that further work on the subscale engine was deemed unnecessary once the cryo deal was signed with the Soviet Union, and was revived only after the Russian refusal to provide the technology.

The regeneratively cooled subscale engine incorporated important technology elements developed entirely indigenously, demonstrating the capabilities of Indian engineers. The coolant channels through which one of the propellants would pass around the engine were made through a process of electroforming. Since electroformed structures can be fabricated as a single piece, weak spots left by techniques such as welding and brazing were avoided, pointed out an article in a 1993 issue of *Propulsion Today*. Electroforming had been used to create the coolant channels of many cryogenic engines, including the French HM7, the space shuttle main engine of the Americans as well as the Japanese LE-7 engine, the article added.

The electroforming technique used by LPSC has been developed by the Central Electrochemical Research Institute (CECRI) at Karaikudi in Tamil Nadu. For the subscale engine, LPSC's machining facility cut channels of varying depth and width in the inner shell of the combustion chamber, which was made of a special high-purity grade of copper called Oxygen Free High Conductivity (OFHC) copper. These channels were filled with wax and made conductive. A process of electrolysis then deposited a layer of nickel over the inner chamber, including the wax-filled channels. After the electroforming, the wax could be melted and removed.

The 30 January 1997 issue of *CSIR News* stated that CECRI had completed a project for the copper electroforming of the Cryogenic Upper Stage (CUS) main engine and steering engine thrust chambers. Instead of nickel, copper would be deposited in this process to form the outer shell. Despite this, the CUS engine will retain the Russian vacuum brazing technique for which a mammoth electricity-guzzling facility has been established.

The regeneratively cooled subscale engine also incorporated an indigenously developed electrical igniter. Unlike the UDMH-N_2O_4 combination, used in the Viking/ Vikas engine, which are hypergolic and spontaneously burn when brought into contact with one another, LOX-LH2 requires a spark to set it alight, rather like the spark plug in a car engine. While a tiny spark would ignite the petrol-air mixture in a car engine, the igniter for a rocket engine needed to generate substantially more heat, commented an article in a 1994 issue of *Propulsion Today*. In the electrical igniter, gaseous oxygen is preheated with an electrical spark and then gaseous hydrogen is injected into the heated oxygen to start the burning.

An electrical igniter is especially advantageous for an engine with multistart capability, i.e., one which can be stopped and restarted several times. But, according to the 1994 talk by Muthunayagam, the indigenous CUS will be using a pyrogen ignition system. Presumably, the imported Russian engine does too. Pyrogen ignition depends on setting off incendiary mixtures. Since the CUS and Russian engines are supposed to have multistart capability, a separate pyrotechnic charge will be needed for each restart. One reason for not using the electrical igniter in the CUS might be that ISRO did not want to risk making alterations to the Russian design.

Likewise, the C-12 engine, which had been part of the GSLV when the original project was approved in 1990, had been sidelined once the contract with the Soviets came through. But some work seems nevertheless to have continued, although at a low ebb. The *1991-92 Annual Report*, for instance, speaks of the detailed design of the 12 tonne cryo engine and stage being taken up.

The indigenous efforts seem to have got a boost after the American sanctions were announced in May 1992 and Russian withdrawal from its contractual obligations became a possibility. There is reason to think that a detailed assessment for the development of the C-12 engine was prepared soon after the US sanctions were announced in 1992. The April-June 1992 issue of *Propulsion Today* carried an item that the gas generator, which would serve as the power source for the C-12's turbopumps, had been successfully tested. The *Performance Budget for 1993-94* says that the configuration of the engine had been finalized. The *1993-94 Annual Report* added that 'work on the development of the indigenous C-12 cryogenic engine has picked up momentum'.

Electroforming was used to create the regeneratively cooled combustion chamber for the C-12. The 1993 article on 'Electroforming Technology' in *Propulsion Today* shows pictures of the C-12 engine's thrust combustion chamber with the coolant channels milled into it and of the chamber after nickel electroforming. The C-12 engine was tested at about a fourth of its rated thrust for about a minute in February 1998.

Although the C-12 was not picked for the CUS development, the technology developed for this engine is likely to be used in the C-20 class cryogenic engine and stage which ISRO plans to develop. When the C-20 configuration is finalized, it will be interesting to see how much of the CUS technology it incorporates. The present indications are that key Russian technology elements which the CUS incorporates, such as staged combustion cycle and vacuum brazing, may not be there in the C-20. If so, it will doubtless raise questions about what ISRO has gained by modelling its CUS on the Russian engine and stage.

The cost of ISRO's cyrogenic decisions

When the GSLV project was sanctioned in November 1990, its aim was defined as the capability to put a 2,500 kg class satellite into GTO. The project was sanctioned with an indigenous C-12 cryogenic stage. The deal with the Soviets for cryo technology, which was signed in January 1991, was supposed to reduce the time for developing the technology, without any increase in the cost. But that is not what has happened.

In the Department of Space's yearly performance budgets, the cost of the Soviet/Russian contract is included as part of the GSLV project costs. But after the Russians refused to transfer the cryo technology, a separate Cryogenic Upper Stage project had to be funded. So the cost of the path ISRO has taken to reach 2,000 to 2,500 kg in GTO has to include both the GSLV project cost and the need to fund the CUS project. For reasons explained earlier, the cost of the Russian contract has increased in rupee terms and contributed to the burgeoning of the GSLV project costs. The combined effect of increasing GSLV project costs and the need to support a separate CUS project are shown in Graph II.

The GSLV project was sanctioned in 1990 with an estimated cost of Rs 756 crore. By 1999-2000, the GSLV and the CUS project together needed a funding of Rs 1,441 crore. That is an increase of nearly 91 per cent, almost a doubling in the cost. By contrast, the cost of the PSLV project increased by just 33 per cent. More important, the GSLVs produced after spending this money will be grossly inadequate to meet the challenges of the present decade (2000 to 2010).

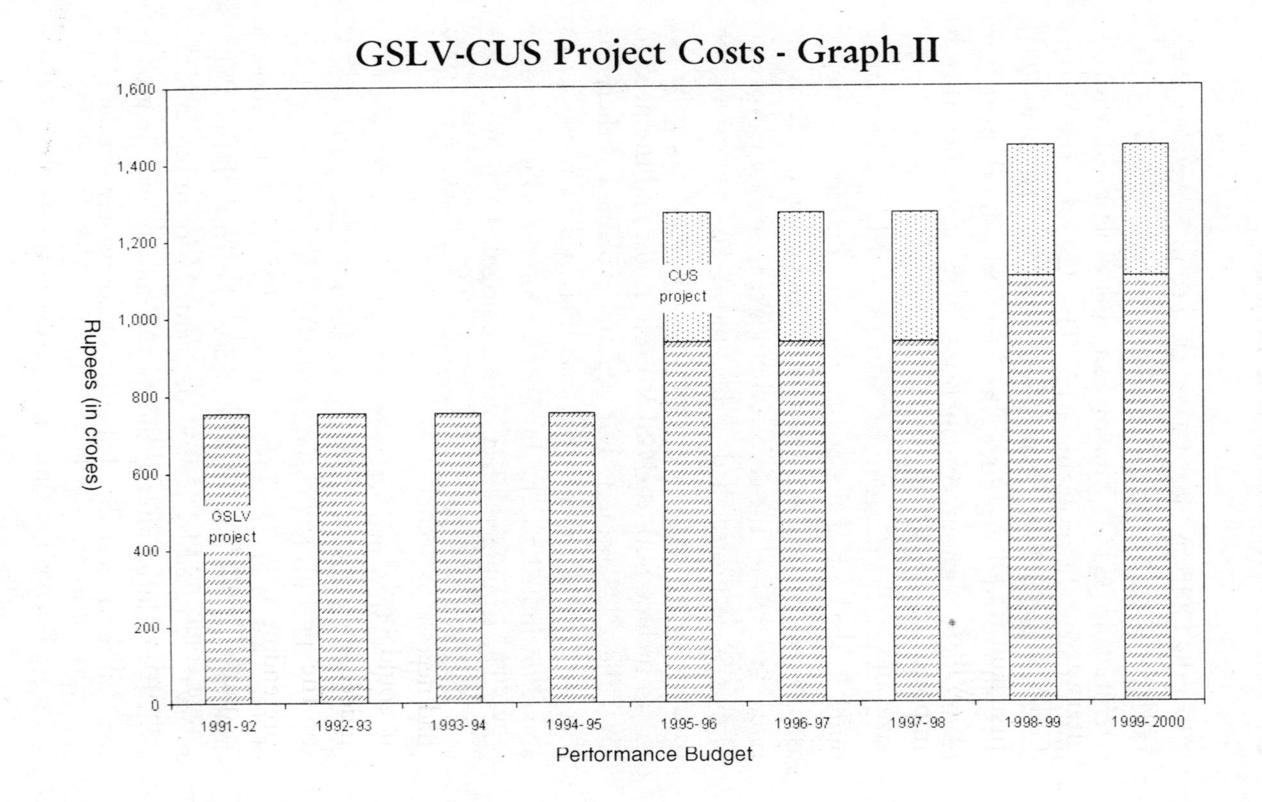
GSLV-CUS Project Costs - Graph II
Rupees (in crores)
1,600
1,400
1,200
1,000
800
600
400
200
0
CUS project
GSLV project
1991-92
1992-93
1993-94
1994-95
1995-96
1996-97
1997-98
1998-99
1999-2000
Performance Budget

When the Cryogenic Study Team submitted its voluminous report in the early Eighties, it spoke of being able to put at least 2,000 kg satellites into GTO. This capability is what the GSLV will now provide in the coming years. When the GSLV project was approved in 1990 and afterwards when the deal with the Soviets was signed, the GSLV was supposed to deliver 2,500 kg in GTO and the first launch of the GSLV was scheduled for 1995.

While the GSLVs, including the Mark-I with the imported Russian stage, have been plagued by delays, the size and weight of communication satellites have steadily grown. Euroconsult, the well-known European space consultancy firm, pointed out in a 1993 report that the 500-800 kg satellites, which accounted for a third of those launched in 1984, had disappeared by 1988. The 1980-91 period was dominated by satellites in the 800-1500 kg class. Almost half of the satellites scheduled for launch from 1992-96 weighed over 2,500 kg. By the mid-Nineties, Arianespace, the world's leading commercial launch company, was looking at average satellite weights in excess of 3,000 kg by the end of the decade.

When the Russians backed off from the deal and ISRO had necessarily to develop cryogenic technology on its own, it could have examined whether the cryogenic stage, and indeed the GSLV configuration, would meet requirements of the period when it would become operational. By pretending that much of the technology had already been acquired from the Russians and that an indigenous cryo stage could be produced with only marginal delays, ISRO passed up an opportunity to turn a misfortune to its advantage.

Driven by the needs of larger capacity, greater onboard power, longer life and more flexibility in using satellite capacity, satellites have continued to become heavier. In its newsletter of April 1998, Arianespace said that spacecraft weighing over 4,000 kg were already on its order books. It predicated that, by 2005, half the satellites launched would weigh over 4,000 kg. At a press conference in Bangalore in February 2000, a senior executive of Arianespace said that the company was already having discussions for the launch of communication satellites weighing 6 tonnes each.

In this environment, the GSLV Mark-I with the Russian cryo stage will be able to carry 1,600 kg to 1,800 kg to GTO. The GSLV Mark-II with the indigenous cryo stage, currently scheduled for launch in 2002, is expected to increase that capability to about 2,000 kg in GTO.

It was only in the *Performance Budget for 1999-2000* that GSLV Mark-III was officially announced for the first time. The GSLV Mark-III will have a very different configuration from the Mark-I and II. According to ISRO officials, the first stage would have clustered Vikas engines. It would be flanked by two large solid strap-ons called S-200, each carrying 200 tonnes or more of propellant. At present, the solid first stage of the PSLV carries 139 tonnes of propellant. The Mark-III would also have a new cryogenic stage, the C-20, carrying 20 tonnes or more of propellant. Although the *Performance Budget for 1999-2000* put the Mark-III's payload capability at 3,000 to 3,500 kg in GTO, the Mark-III will probably have a payload capability of at least 4,000 kg in GTO by the time its configuration is frozen and it is formally approved as a project.

The GSLV Mark-III has not yet been approved as a project. The *Performance Budget for 1999-2000* included just preliminary funding for the GSLV Mark-III, the S-200 and the C-20.

The private assessment given by ISRO engineers is that since the Mark-III is a completely new vehicle, its development would take eight years at the very least, but more like ten years. This matches the ISRO experience: the SLV-3, the PSLV and the GSLV (even with the imported cryo stage) have taken around ten years to develop.

The Ariane experience also suggests that eight to ten years is a realistic time-frame. It took nine years to go from Ariane-1's 1,800 kg in GTO to the 4,900 kg capability of the Ariane-4. Ariane-4 had the advantage that all its basic technology elements, such as the Viking engines, the clustered liquid first stage and the cryogenic upper stage, were present in the Ariane-1 and could be upgraded through the intermediate Ariane-2/3.

If the Mark-III does indeed take about ten years, then it would leave the GSLVs with the capability to put only around 2,000 kg into GTO all through the 2000-10 decade.

The 2000-10 period is likely to be most challenging decade which ISRO has faced so far, especially in the highly-profitable communications segment. In the Nineties, ISRO had the satisfaction of watching its transponders being rapidly put to use as the indigenous Insat-2 satellites were launched in quick succession. But in the coming decade ISRO will no longer have the advantage of being the monopoly supplier of Indian communication satellites. A Cabinet decision in 1997 approved a satellite communication policy framework, allowing private companies to establish commercial communication satellites. Another Cabinet decision in January 2000 allows Indian companies with up to 74 per cent foreign equity to establish and operate satellite systems.

Communication and broadcast services, and therefore utilization of satellite capacity, which used to be entirely a government monopoly, is steadily being opened up and exposed to competition. Private TV channels broadcast by non-ISRO satellites have already become entrenched. Private internet service providers have made their appearance. Private cellular operators are present in all cities and even in some of the bigger towns. Private companies are already providing high-speed data links over satellites using VSATs (Very Small Aperture Terminals). The direction of change is clearly towards allowing private companies to provide basic telecom services and long-distance telecommunication traffic.

The upshot of all this activity is going to be that private companies as well as their government competitors will be looking for the best deals from satellite builders and operators. So ISRO's satellites will have to be competitive in the marketplace. If 4,000 kg (or heavier) communication satellites are the most cost-effective solution, then that is what ISRO will have to build and launch. ISRO's satellite designers will need to gear up to produce heavier satellites. If the GSLV is not able to launch them, then these satellites will be launched by foreign launch vehicles, as the Insats are today.

After PSLV's first successful flight, only one IRS satellite (the IRS-1C) was launched abroad because it was too heavy. The PSLV was rapidly upgraded and could launch the IRS-1D, identical to the 1C, just two years later. If ISRO's communication satellites continue to be launched abroad after the GSLVs become operational, it could raise fundamental questions about the relevance of the GSLV programme.

ISRO's assessment

Although the ISRO chairman, K. Kasturirangan, agrees that the ability to launch larger and heavier communication satellites is required, he takes the view that the GSLV Mark-

III can be got ready in five or six years. In the meantime, the current GSLVs can be used to launch 2,000 kg satellites which will offer novel and economical solutions. In an interview for this book, he said:

> There will be pressure to go in for bigger communication satellites as the price per transponder per year comes down only when a satellite carries 24-36 transponders and these transponders are of high power. So the GSLV Mark-III, which can launch 4-4.5 tonnes in GTO, is being developed. In terms of capability, it will be comparable to the best launch vehicles currently flying. The GSLV Mark-III can be done in the next five or six years.
>
> In the meantime, ISRO's strategy is to launch 2,000 kg class communication satellites, carrying very innovative payloads, using the current GSLVs. The combination of such satellites being launched by the GSLVs will provide economically attractive solutions and permit a smooth transition to a world-class system by 2007-08.
>
> ISRO is developing a 2,000 kg class of communication satellites which will use a common basic structure. These are intended to be workhorse communication satellites, each carrying about twenty-four transponders. Even if the size of these satellites is on the smaller side, it is hoped that they can be produced and launched in sufficient numbers to make them cost-effective.
>
> These satellites will have better frequency spectrum utilization, more onboard processing for various purposes, greater flexibility through concepts such as dynamic switching of power depending on the traffic and so forth. These advanced technology satellites will start being launched three to four years from now.
>
> Apart from the space segment, there has been considerable investment in setting up a large ground

infrastructure tuned to using the present transponder technology. These ground systems too will need seven to eight years for transition to newer technologies.

I feel that the GSLV Mark-III can be got ready in the next five or six years because there are no major uncertainities in the technology involved. We are taking a minimal risk route and will stick to upgrading existing technological capabilities. We are not undertaking development of anything completely new and untried. The large solid motors have to be scaled up, the liquid motors have to be clustered and there will be a more powerful cryogenic stage. In the SLV-3, ASLV and even the PSLV, it was the uncertainities in technology which prolonged their development.

In the GSLV Mark-III, we are reducing the number of stages. Apart from the solid strap-ons, it will have two stages. Reducing the number of stages increases the reliability of the launch vehicle. Almost half the velocity needed to put the satellite into orbit will be achieved by the time the booster stage burns out and the cryo stage will provide the remaining velocity required. This elegant solution also takes care of the range safety problems affecting the current GSLVs.

In the Mark-III, the number of components required for the liquid stages and indeed the entire launch vehicle will be reduced. Each component requires time for manufacture, assembly and testing, and therefore contributes to cost. Optimization of the components will add to the launcher's reliability too.

Going entirely indigenous from the beginning

Faced with the need for cryogenic technology in the early Eighties, ISRO assumed that it could cut short development time and risks by buying the technology from abroad. Failing to take into account an increasingly hostile international geopolitical environment, ISRO went about decision making in a remarkably leisurely fashion. By 1983-84, the need for cryogenic technology was quite evident and a detailed study report examining how to develop the technology had already been completed. But for seven years ISRO dithered over whether to import the technology or develop it indigenously.

ISRO's decision to import cryogenic technology was questionable, considering the international milieu and the consequences for the launch vehicle programme. Such a high-risk strategy would have been understandable if there was no viable indigenous option open to it. Indeed, if ISRO had strongly pushed indigenous cryogenic development right from the beginning, a fully indigenous GSLV could have been flying by now.

The argument has been put forward that ISRO's liquid propulsion group could not have embarked on a full-scale cryogenic engine immediately after the *Cryogenic Systems Studies* report was submitted as it had its hands full getting the Vikas engine built and in developing the PSLV's second and fourth stages. The strange paradox is that some senior people in liquid propulsion say that ISRO ought to have leapt at the chance when the French reportedly offered the technology for their HM7 cryogenic engine at a low price. Even before it had proved that it could build the Viking engine and a suitable stage in India, the liquid propulsion group was ready for yet more technology acquisition from France. It does not appear to have lacked people for such a venture.

It is noteworthy that the liquid propulsion group did not carry out any experimentation with semi-cryogenic or cryogenic technology, even after the Kurup Committee pointed out the importance of doing so. At the very least, immediately after the *Cryogenic Systems Studies* report was submitted, preliminary steps, such as the subscale engine development, could have been initiated. The development of the engine and stage could then have begun in earnest a couple of years later. China, another developing country, made its first cryogenic engine in eight years. Even if ISRO had taken ten years, an indigenous cryogenic stage would be ready by now.

ISRO's problems with cryogenic technology have not been due to lack of technical capability but arose out of poor management decisions. The liquid propulsion group's import culture, which took deep root after the Viking experience, contributed in no small measure to this.

CHAPTER

10

The Past, the Present and the Future

THE INDIAN SPACE programme and its architect, the Indian Space Research Organization, have been remarkably successful. The services in communication, broadcasting, and meteorology are now being provided by satellites designed and built in India. Despite its unorthodox configuration, the PSLV has become a rugged and reliable operational launcher.

Among developing countries, only China and India have successful space programmes, including the capability to launch their own satellites. Most other countries with satellite launch capabilities, including China, were able to draw upon the experience of their military ballistic missiles and launchers in doing so. Japan and India have been the only two exceptions. The Indian launch vehicle programme was begun in unpropitious circumstances. It had to cope with the problems of economic underdevelopment, bringing with it issues of securing funding and limited industrial infrastructure. In addition, there was no prior experience in rocketry which it could draw upon.

The achievements of the Indian space programme and the fact it was managed on relatively small budgets has been recognized only in recent years. 'Despite its limited resources, India has and is continuing to develop a broad-based space program with indigenous launch vehicles, satellites, control facilities, and data processing,' says an analysis prepared by the Federation of American Scientists. 'Since its first satellite was orbited by the USSR in 1975 and its first domestic space launch was conducted in 1980, India has become a true space-faring nation and an example to other Eurasian countries wishing to move into the space age.'

What, then, have been the factors which have contributed to this success?

Focus on applications

The Indian space programme's greatest asset has undoubtedly been Sarabhai's vision. In the early and mid-Sixties, when applications using satellites were still experimental even in the United States, Sarabhai was quick to recognize their benefits for India. He foresaw that satellites could usefully supplement ground-based systems for providing many services in communications, direct TV broadcasting, remote sensing and meteorology.

The Indian space programme was, therefore, application-driven from the very beginning. Building satellites and launch vehicles were a means to that end, not an end in themselves. But Sarabhai did believe that the very process of engineering complex systems involving ground equipment, satellites and launch vehicles would establish technical capabilities as well as multi-disciplinary teamwork. He thought that such capabilities too were essential for India's progress.

With such a holistic vision right from the beginning, ISRO was able to set its priorities and use its limited budget to the best advantage. There was a clear step-by-step strategy. ISRO would not wait for its own satellites to begin application development. Applications would be proven, ground systems put in place and users familiarized with the new technology by using foreign satellites. Thus when the indigenous satellites became available, they could immediately be utilized to the full. Since indigenous launch vehicles to put these operational satellites would take longer to develop, the satellites would initially be launched from abroad. As the PSLV experience shows, when the indigenous launcher became available, it was speedily put to use.

In remote sensing, an aerial survey using infra-red film was carried out from a helicopter in 1970 to study root-wilt in Kerala. An infrared scanner had been developed in association with France. It was brought to India in 1972 and tested for various applications. Even before the Americans had launched the world's first civilian remote sensing satellite, the Earth Resources Technology Satellite (later renamed as Landsat), ISRO had requested access to data from the satellite. As a result, India became one of the earliest users of Landsat when the first satellite was launched in mid-1972. Subsequently, a ground station to receive Landsat data directly from the satellite was set up and India became a major user of Landsat remote sensing data. Landsat data was used to prove applications and develop users, especially among other government departments and agencies. Thus, by the time India's first remote sensing satellite, the IRS-1A, was launched in March 1988, there was no need to prove the usefulness of the satellite or look around for users.

Sarabhai realized that communications and direct TV broadcasting by satellite was particularly advantageous for a country as vast as India whose ground infrastructure to provide these services was ill-developed. As early as 1967, ISRO conducted a joint study with NASA which showed that a hybrid system of direct broadcast by satellite combined with terrestrial TV transmitters would be the most cost-effective method for nationwide TV coverage. In September 1969, ISRO signed an agreement with NASA for the Satellite Instructional Television Experiment (SITE) experiment. As a result, NASA lent its ATS-6 satellite to India for a year, during 1975-76, to demonstrate direct TV broadcasting as an instructional medium for reaching India's numerous villages. SITE became the earliest large-scale experiment with direct broadcasting anywhere in the world.

In 1975, ISRO signed an agreement to use the Franco-German satellite Symphonie for the Satellite Telecommunication Experiments Project (STEP). One of Symphonie's two tranponders was made available to India from June 1977 for experiments with communications over satellite, radio networking and TV transmission.

ISRO conducted a joint study with the Massachusetts Institute of Technology in 1970 on the design of the Indian National Satellite (Insat) and the government approved the establishment of the Insat-1 system in 1977. These multipurpose satellites were intended to handle communications and broadcasting and each satellite also carried a meteorological camera. They were built by Ford

Aerospace (now part of Space Systems/Loral) and launched abroad. Although only two of the four Insat-1 satellites, the Insat-1B and the 1D, gave sustained service, together they provided satellite services from 1983 onwards. As in remote sensing, users and ground equipment were put in place with these satellites. There was a smooth transition when the indigenous Insat-2 satellites were launched from 1992 onwards.

While foreign satellites were used to create users and the necessary ground infrastructure, ISRO first built experimental satellites and then operational ones. The first Indian satellite, the Aryabhata, was launched from the Soviet Union in 1975. The Bhaskara-I and Bhaskara-II, launched from the Soviet Union in 1979 and 1981 respectively, were the first attempt to build earth observation satellites. These satellites, along with the Rohini satellites launched on the SLV-3, were the precursors for the operational Indian Remote Sensing (IRS) series of satellites. The Ariane Passenger Payload Experiment (APPLE), carried on the Ariane's third launch in June 1981, similarly provided the hands-on experience needed to build the second-generation Insat communication satellites within the country.

As earlier chapters have shown, ISRO followed an equally systematic step-by-step process, starting with sounding rockets, for developing the capability to put its operational satellites into orbit.

The result of this strategy was that benefits from the space programme began to flow quite early, demonstrating that claims made about the usefulness of satellites were not just talk. Using data from the US Landsat satellite, remote sensing applications addressing real-life problems began appearing from the mid-Seventies. The SITE and STEP experiments of the mid-Seventies set the stage for operational communications and broadcasting services with the foreign-built Insat-1 satellites from the early Eighties. From the late Eighties, India's own operational satellites began to appear. The first IRS was launched in 1988 and the first indigenous Insat in 1992. In October 1994, a long-cherished dream materialized for the first time, an Indian launch vehicle, the PSLV, carried an indigenously built operational satellite, the IRS-P2, into orbit.

This strategy recognized that launch vehicles with the capability to put operational satellites into orbit would take considerable time to develop. By getting applications operational as quickly as possible and then replacing foreign satellites with indigenous ones, ISRO has been able to secure sustained and growing government support for nearly four decades now. The purpose for which the launch vehicles were being built was also clear.

Space and defence programmes

Western analysts usually make much of the links between India's launch vehicle programe and its missile programme. Unlike the Department of Atomic Energy, which has undertaken the development of nuclear power and other peaceful uses of atomic energy as well as the development of atomic bombs, the Department of Space has restricted itself to developing space applications, satellites and launch vehicles. Missile development has been the responsibility of the Defence Research & Development Organization (DRDO). The missile programme has certainly benefited from technology, expertise and facilities developed in the course of the space programme, most notably in the case of the Agni long-range missile. But then every country uses all the resources at its disposal to counter threats to its security and there is no reason why India should be an exception.

It is, therefore, all the more important to recognize that the course of the space programme and the choice of technologies, including those for the launch vehicle component of the programme, has been shaped by ISRO's own goals, needs and perceptions, not those of the country's strategic programmes. To give just one example, ISRO's reliance on solid propulsion has often been interpreted in the West as India using its launch vehicle programme to develop critical technology needed for its missile programme. Use of the SLV-3 first stage in the Agni missile is held up as proof of this.

But, as narrated earlier in this book, ISRO's use of solid propulsion came about as a result of historical factors. In the early days of the launch vehicle programme when technological capabilities had to be built from scratch in the face of limited funding and poor industrial infrastructure, solid propulsion probably appeared the simpler, quicker and cheaper route. The decision by Japan to opt for the solid route for its first launch vehicle would also have counted, particularly since Professor Hideo Itokawa of Japan's Institute of Space and Aeronautical Sciences was a consultant to ISRO.

Once solid propulsion capability was created, it generated its own momentum, aided by the fact that ISRO's solid propulsion groups were successful and dynamic. Nevertheless, despite the 'solid propulsion lobby' which ISRO's liquid engineers often complain bitterly about, the use of liquid engines has steadily grown. While the SLV-3 and ASLV were entirely solid propelled, with liquid engines only for attitude correction, half of PSLV's four core stages were liquid. In the GSLV Mark-I and II, only the first stage inherited from the PSLV will be solid. The GSLV Mark-III will probably have a configuration typical of launch vehicles in other countries: all the core stages being liquid, with solid strap-ons providing additional thrust at lift-off.

Strangely, if ISRO was biased towards solids, it was DRDO's missile programme which favoured liquids! The result is that the Prithvi missile is liquid propelled, using liquid engines reverse engineered from a Soviet missile. Even the Agni-I missile had a liquid second stage using the Prithvi-type liquid engine. The Agni-II, with both stages solid, was tested for the first time only in April 1999, ten years after the first flight of the Agni-I. Since missiles need to be launched at short notice, it would have made sense to have Prithvi and Agni as fully solid missiles. Having successfully launched the all-solid four-stage SLV-3, ISRO was certainly capable of developing such solid missiles. The fact that ISRO's solid propulsion capability was not utilized in this fashion shows that it was created to meet ISRO's needs and not those of the missile programme.

Moreover, the Brazilian experience is perhaps a salutary lesson in what can happen when a developing country's launch vehicle programme becomes too closely associated with missiles. The Brazilians, like their Indian counterparts, developed and launched indigenous sounding rockets, the Sonda series, in the mid-Sixties. But in the face of Western technology embargoes (which ISRO too had to face) and wavering financial support for the programme, Brazil has yet to successfully fly its first launch vehicle, the VLS. By contrast, the Indian space programme received sustained public support and funding precisely because it was so strongly rooted in applications which were demonstrably beneficial for the country.

Managing the geopolitics of space

Sarabhai had recognized right from the beginning that launch vehicle development would be complicated by its 'military overtones'. Much of the technology which goes into a launch vehicle can, after all, be used for long-range missiles as well. The United States, the then Soviet Union, and China were able to convert their liquid-fuelled ballistic missiles into space launchers.

There is no logical reason why technology cannot flow in the reverse direction as well — from civilian launchers to long-range missiles — and the launch of the SLV-3 had a considerable influence on the establishment of the Missile Technology Control Regime (MTCR) in 1987. The MTCR's goal was to control transfers which would contribute to delivery systems (other than manned aircraft) capable of carrying nuclear warheads and subsequently its ambit was increased to cover chemical and biological warheads as well. The MTCR's greatest flaw, certainly from the Indian perspective, has been that it was a continuation of the Nuclear Club philosophy. Its purpose was solely to prevent the technology from going to non-club members, without in any way restraining the use of such technology by the club members. The United States was, for instance, free to sell nuclear-tipped ballistic missiles to Britain and import rocket engines from Russia.

But it is not as if launch vehicle technology was available for the asking before the MTCR. As has been recounted in this book, the United States refused to help ISRO set up the Static Test and Evaluation Complex (STEX) at Sriharikota for the testing of large solid motors. Likewise, although ISRO could get sample quantities of the solid propellant resin CTPB, it was unable to purchase it in bulk.

After the launch of the SLV-3, ISRO could not buy the PBAN propellant resin used in the launcher's lower two stages.

ISRO had always been aware that its launch vehicle programme would invite international suspicion. The aggressive indigenization it followed from the start was able to limit the programme's vulnerability to technology embargoes. Bulk materials such as solid propellant ingredients, liquid propellants and maraging steel were obvious priorities, as also were inertial sensors.

But for this early emphasis on indigenization, it is quite possible that the launch vehicle programme would have been in serious trouble, perhaps even halted, by the technology denial regimes. Since the Eighties, there has been steady tightening of export controls over 'dual use technologies', including launch vehicle technology. France, which gave the Viking technology to India, later refused to supply the silica phenolic throat material needed for the Vikas engine. In mid-1992 the US imposed sanctions on ISRO over the cryogenic deal with the Soviet Union. A few years back, Krupps refused to roll the maraging steel rings needed for the PSLV first stage. Since ISRO always had indigenous or other options available to it, the launch vehicle programme was unaffected.

The ability to foresee and handle such geopolitical realities has therefore contributed much to the success of ISRO's launch vehicle programme. The cryogenic deal with the Soviet Union stands out as the one time when ISRO let its guard down. Despite the existence of technology control regimes, US laws to enforce these export controls, and having an indigenous development option open to it, ISRO decided unwisely to go ahead with the import of technology. The launch vehicle programme has paid a heavy price for that decision.

US restrictions on securing international launch contracts is a problem which ISRO could face in the future, especially if it tries to market the GSLV. Something like 70 per cent of the world's civilian geostationary communication satellites are built in the United States. As US export controls extend to critical satellite components as well, it is able to influence the launch of even non-US-made satellites.The US was able to place quotas on the commercial launches which Russia, Ukraine and China could accept as well as limits on how low a price they could charge. The quotas for Russia and Ukraine were removed only in the year 2000. But growing concern in the US that launches given to China could help that country improve its nuclear missiles has led to tightening of US export controls on satellites and satellite components. Since the US and India disagree on many non-proliferation issues, similar concerns and problems could occur when ISRO tries for launch contracts abroad.

Resource allocation linked to delivery

Graph III shows marked phases of growth in ISRO's annual budget, suggesting that funding increases came only after the organization demonstrated its capabilities and benefits from space activities. In the graph, ISRO's yearly budget allocations from the government have been converted to present-day prices to make the comparison more meaningful. So the years when ISRO's budget appears to have declined indicate periods when its budget increases did not keep pace with inflation.

In the Seventies, when the the SLV-3 and the early experimental satellites were being developed, ISRO's average

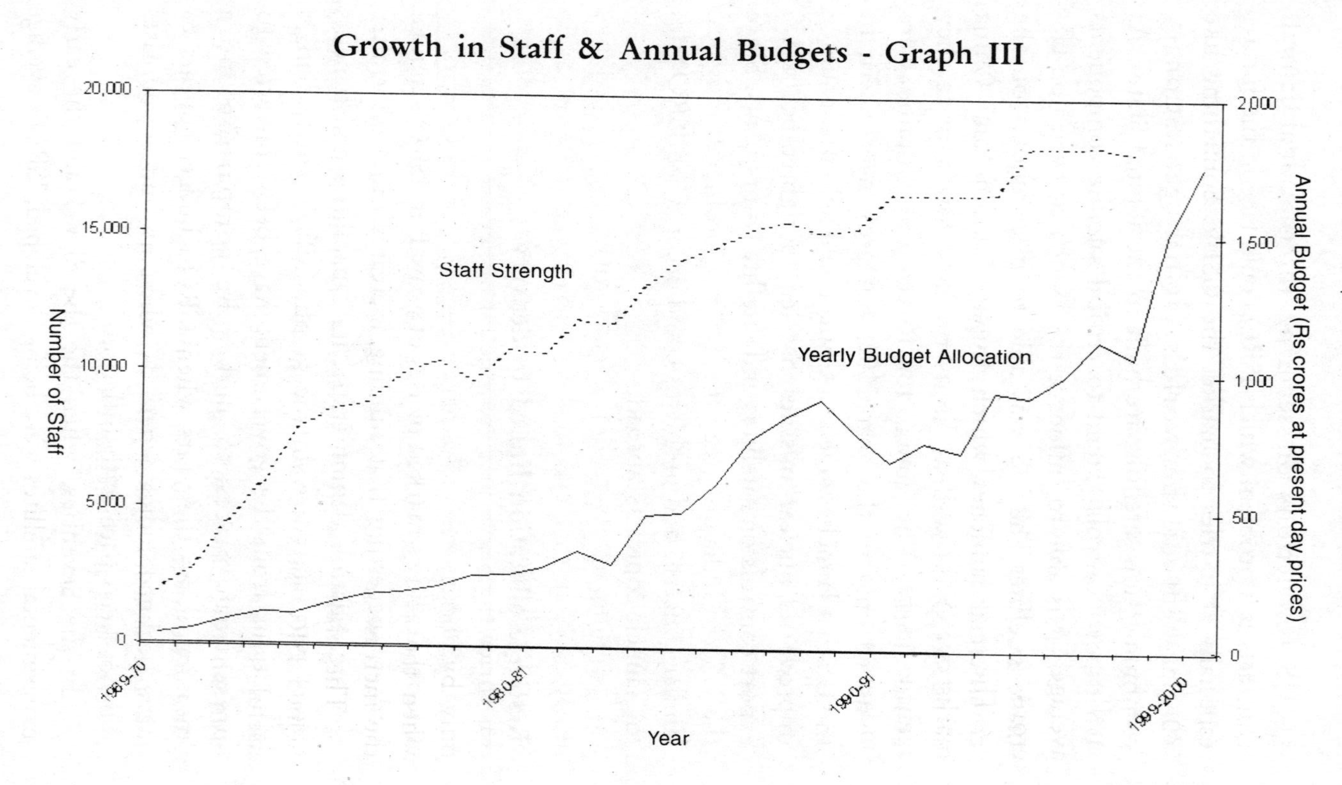
Growth in Staff & Annual Budgets - Graph III
Number of Staff
20,000
15,000
10,000
5,000
0
Staff Strength
Yearly Budget Allocation
Annual Budget (Rs crores at present day prices)
2,000
1,500
1,000
500
0
1969-70
1980-81
1990-91
1999-2000
Year

yearly budget was only about Rs 165 crore. During the Eighties, when the experimental phase was successfully completed and ISRO began developing operational launch vehicles and satellites, its average yearly budget tripled to over Rs 550 crore. In the Nineties, when these operational satellites and launch vehicles were entering service, the average yearly budget further doubled to over Rs 1,000 crore.

The graph also highlights another characteristic which marks out ISRO. Staffing in government departments and agencies in India generally tend to grow along with their budgets. As a result, salaries and other administrative expenses swallow much of the budget increases. Consequently, funding for improvements in function and operations gets curtailed.

The graph clearly indicates that ISRO's increases in staffing have been related to its needs, not to funding levels. The staff strength increased most when the organization was building up its technical capability. It increased by nearly 5,000 in the Seventies when the programme got into its stride. In the Eighties, when the average annual budget was thrice the Seventies level, the number of staff increased only by about one and a half times. But in the Nineties, when the average budget was more than Rs 1,000 crore, the increase in staff has been less than 2,000.

The ability to hold down its internal expenses has helped ISRO become a successful organization, providing useful services and products. A key ingredient in this has been an early and consistently growing commitment to using industry as much as possible, rather than build up in-house capacity. Industrial participation will be discussed in detail later in this chapter.

The management culture

ISRO's success owes much to the effectiveness of its management methods, which seek to minimize bureaucratic interference in technical decision making and increase coordination within the organization as well as with industry.

The space programme had the advantage of being modelled on the atomic energy programme. As in the latter, the space programme is always headed by a technical person and not, as in most other government departments, by a career bureaucrat from the Indian Administrative Service. These days, it is a senior person from within the programme itself. This person remains the head of the programme till he retires from government service. As a result, both atomic energy and space enjoy continuity in leadership which is denied to other departments, whose heads are shuffled around every now and then.

Again, as in atomic energy, the formal administrative structure of the space programme consists of the Department of Space (DOS) and the Space Commission. The DOS is an arm of the Indian government and is able to exercise all the powers of a government department. The Space Commission has, in the words of the government resolution which set it up, 'full executive and financial powers modelled on the lines of the Atomic Energy Commission'. All important programmes and projects have to be cleared by the Space Commission. No separate approval is then necessary from the Ministry of Finance. But approval of the Cabinet may be needed for high-expenditure projects and clearance of the Planning Commission required for incorporation in the Five Year Plans. Similarly, it is the Space Commission, not the

Ministry of Finance, which clears the department's budget requests for inclusion in the Union government's annual budget.

It was Sarabhai who set up the Indian Space Research Organization through a departmental order. Space activities had been coordinated in the initial years by the Indian National Committee for Space Research (INCOSPAR). After INCOSPAR was reconstituted under the Indian National Science Academy, Sarabhai created ISRO in 1969. When the space programme was separated from atomic energy in 1972, ISRO was brought under the newly created Space Commission and the Department of Space.

The creation of ISRO made it possible to have some degree of separation between the administrative functions of a government department and technical operations. Much greater technical inputs and coordination are needed for a high-technology programme which could not have been achieved with the usual hierarchical structure of a typical government department. The ISRO Council brought together the directors of the various ISRO centres as well as senior officers of the Department of Space. This body has 'provided a symbolic link as well as a forum for participative management between the department which has the government's powers and the centres which execute the jobs,' points out Y.S. Rajan, former ISRO scientific secretary. 'It should be noted that in the normal parlance of traditional government departments, ISRO's centres and units would be "subordinate" or "attached offices", words which are not, however, prevalent in ISRO and DOS. Such participative management, jointly evolved between those who wield administrative powers and the executing agencies, is an important feature of the management of the Indian space programme,' he adds.

Satish Dhawan established the ISRO headquarters to assist him in technical decision making. The headquarters staff is made up of technical personnel and is headed by the ISRO scientific secretary. The most important function of the headquarters staff was systems analysis and planning, according to Dhawan. The headquarters staff was expected to look at the components which made up the space programme and analyse options for their coordinated growth. Sarabhai had strongly believed that systems analysis and planning were the bedrock for the development of a scientific programme, says Dhawan.

While the Space Commission and the Department of Space have legally defined powers, the ISRO council and ISRO headquarters only exercise powers delegated to them by the head of the space programme. Nevertheless, ISRO has become, as official documents such as the Performance Budget phrase it, 'the primary agency charged with the responsibility of executing the research and development programmes and schemes of the Department in accordance with the directives and policies laid down by the Space Commission and the Department of Space'. So much so that the head of the space programme is best known as chairman of ISRO, although he is also chairman of the Space Commission and secretary for the Department of Space.

Atomic Energy, and then Space, made full use of the freedom given to them not to follow the usual government procedures. In Space, powers, including the all-important one of incurring fairly substantial levels of expenditure, were extensively delegated to the project directors, the project management boards, centre directors and the ISRO council. The emphasis has been on completing a project on schedule.

The early chapters of this book have illustrated how

Sarabhai had an informal management style. All proposals were presented to him and he would decide, often then and there, which ones could go ahead. In many cases, two or more people and groups could be working on the same problem. It all seems extraordinarily chaotic. But Sarabhai, who had considerable management experience, must have had good reason for choosing such a style.

The most probable reason was that his aim at that stage was to create basic competence in various technology elements which go into rocketry as well as to get people used to working in teams. The young men he recruited had sound technical training, plenty of enthusiasm but, with rare exceptions, no experience in rockets. It would have been difficult to predict how each of them would turn out. Under the circumstances, Sarabhai encouraged people to experiment and fostered competition. By the time he died, he had succeeded in creating sufficient capability and experience to take on the challenge of building a launch vehicle.

Dhawan was then able to put in place the management structures which have served ISRO well. Each project had a project director assisted by a core team of specialists drawn from different groups and ISRO centres. This project management methodology was further improved and refined. The project team usually reports to a project management board made up of the directors of the ISRO centres involved. The project management board performs both review and coordination functions. As stated earlier, the project director and the project management board have powers to sanction expenditure.

Dhawan also introduced two other elements, both common in Western management systems but still relatively novel in the Indian context. One was the system of commissioning studies to look at various options for ISRO's future course of action. The studies done by the Vasagam and Srinivasan Committees, for instance, paved the way for the PSLV project. In other cases, such as the study looking at development of a liquid oxygen-kerosene engine for the PSLV, the issues would have been examined and discussed in depth before a decision was taken not to pursue the option suggested. Irrespective of whether such studies became projects or not, they brought together specialists from various ISRO centres at a very early stage in the planning process. As a result, there was better understanding of each other's problems and concerns. It reinforced the organization's ability to run cohesive multi-disciplinary teams.

The other was a review system for projects. Every project being executed by ISRO passes through regular reviews at various levels. These formal and informal reviews, including by the ISRO chairman, provided a way of quickly resolving conflicts and differences of opinion without undermining the delegation of authority, points out Y.S. Rajan. The preparation for such reviews, which involves formal documentation and presentations, often helped clarify issues. These periodic reviews have also helped ISRO keep to schedules as far as possible and limit cost and time overruns.

Industry as a partner

Right from the beginning, Sarabhai had insisted that the capabilities available in industry be used to the fullest. His successors continued this policy, each taking initiatives to improve industrial capability and increase industry's involvement in the space programme. Industries which could take on space-quality tasks did not exist when the programme began and ISRO's involvement with industry had necessarily to be much deeper, points out P. Sudarsan, who as head of Technology Transfer and Industry Cooperation at ISRO headquarters, played a key role in establishing ISRO's interactions with industry. So the typical government procedure of floating a tender and giving a contract to the lowest bidder simply would not work. Instead, ISRO put in place a system for 'vendor development'. ISRO would make an assessment of the capabilities of industries and help improve those capabilities so that they could undertake tasks to ISRO's stringent specifications. Help from ISRO could take the form of technology, assistance in overcoming difficult problems and putting in place test and quality assurance procedures.

In going to industry on such a scale, ISRO would have had to continually resist pressures to increase in-house production capacity. Nor, like the atomic energy programme, did it establish public sector companies to carry out these functions. It is also remarkable that ISRO chose to involve private industry, rather than confine itself to just public sector companies. Since, as stated earlier, ISRO's involvement with an industry went much deeper than just signing a contract, a government department would usually prefer to deal with public sector companies to avoid allegations of wrongdoing or favouring a company. As each chairman has emphasized industrial participation, ISRO has been able to evolve systems to negotiate contracts with industry, have ways of extending a helping hand when necessary, and at the same time make sure that its delivery schedules and product specifications are met.

In a talk given by Satish Dhawan in 1983, he said that the hope was that the bulk of the investments in the space programme would flow into the Indian economy. Such investments flowing to the Indian industrial and technology sector ought to create a multiplier effect through technology, quality, reliability and skill upgradations. This should bring economic benefit as well as technological strength. 'Active industrial participation not only directly helps the space programme but also is, technologically and in the long run, economically profitable for industry,' he added.

More than forty industries contributed to India's first launch vehicle, the SLV-3. This number had risen to over 150 industries in the PSLV and something like 70 per cent of the budget for the PSLV programme is estimated to be going to industry. In 1992, the total flow of funds to industry from the space programme was Rs 248 crore, close to half its total budget.

Dhawan's efforts to develop a partnership with Indian industry showed his foresight regarding the needs of operational systems in this country, especially when the programme had to be self-reliant, points out Y.S. Rajan. While Dhawan might have derived ideas from the developed countries, a number of unique organizational linkages such as formal memoranda of understanding with major industries and formation of compact groups in ISRO to interface with industries were his own contribution.

There was a quantum jump in industrial involvement when ISRO embarked on developing its first operational launch vehicle, the PSLV. The PSLV was not only a much larger vehicle than the SLV-3 and the ASLV, but also one which ISRO intended to launch regularly. It created the potential for much larger contracts and repeat orders being given to industry. Contracts for bulk supply of materials like HTPB, UDMH-N_2O_4, and maraging steel were negotiated and signed in the mid-Eighties. There were major fabrication tasks as well, such as of the huge first stage motor casing, propellant tanks and gas bottles, various stage and interstage structures, and the heat shield.

U.R. Rao took the important step of moving away from simply fabricating components and parts in industry. Godrej and Machine Tool Aids & Reconditioning (MTAR) received contracts for supplying the entire Vikas engine. Subsequently, the public sector Kerala Hitech Industries Limited (Keltec) received a similar contract. Each PSLV carries one Vikas engine and each GSLV requires five such engines. Industrial involvement was therefore essential for ISRO to make the PSLV and the GSLV operational. These three companies are said to have already supplied over twenty-five Vikas engines to ISRO.

An R&D organization like ISRO was not well suited to carrying out repetitive jobs, points out Rao. At the same time, Godrej and MTAR had to learn to work with one another and subcontract certain tasks to other vendors, instead of buying expensive equipment and establishing the full range of facilities in-house.

Rao has been responsible for another important initiative. In the development of the cryogenic engine, industry was involved right from the beginning rather than have technology transferred to it after the prototype had been built. By doing so, it is hoped that the delays involved in transferring technology can be avoided.

The policy of using industry had helped ISRO retain its role as a high-technology R&D and mission-oriented organization, said Rao in the first issue of ISRO's *Space-Industry News,* which came out in 1989. 'ISRO is able to have access to, by and large, all the requirements for its programmes, without itself getting involved in routine manufacturing activities for which industry is better equipped and capable. The industry, in its turn, has gained financially and also in terms of access to sophisticated, indigenous technology,' he pointed out.

The present chairman, K. Kasturirangan, would like to take industrial participation in the space programme still further. 'My concept is of a very vibrant space programme with investments which are much beyond what the country is currently capable of through the taxpayer's money. We need to have a space industry which is able to produce hardware and provide services not only for India but wherever there is possibility of a good market,' said Kasturirangan in an interview for this book.

At present, industrial participation is only through manufacturing, assembly and some limited testing for ISRO. 'Like Arianespace, a consortium of industry, involving also government and public sector bodies, should be able to come together to produce launch vehicles and sell launch services,' he remarked. ISRO had some unique facilities and a policy decision would be needed to make these available at a price to such a launch vehicle agency.

ISRO's task would be to remain in the forefront of research and development and to provide the latest technology. It should decouple itself from repetitive production and being a service provider. Production of operational systems and provision of services should be the responsibility of operational agencies. But there would have to be a close and symbiotic relationship between ISRO and these operational agencies, points out Kasturirangan. A phased programme was necessary so that industry was able to shoulder increasingly greater responsibilities and also to forge a suitably close and enduring relationship between ISRO and industry.

'We are looking into what operations can be done by industry and what has to be done by us,' says D. Narayanamoorthi, director of the Launch Vehicle Programme Office at ISRO headquarters. The aim is that ultimately industry should be able to produce all the stages and also integrate the full launch vehicle.

As a step in that direction, ISRO would like to see industry deliver some of the stages in ready-to-fly condition. Since the Vikas engine is already being produced outside ISRO, industry is likely to be asked to deliver the fully assembled engine. At present, the final assembly and testing of the Vikas engine is carried out by ISRO. ISRO is also known to be actively examining the possibility of having fully assembled stages with the Vikas engine supplied by industry.

The extent to which ISRO is able to build up a strong space industry could become a key element in the growth and success of the Indian space programme in the years to come. Europe has been successful in using government funding for technology development and having companies carry out the subsequent commercial operations. Arianespace, which was started in 1980, is today the world's leading commercial launch company. Likewise, Spot Image set up by the French has become an important player in the remote-sensing market. A thriving space industry will be able to invest money of its own and give the government more reason to invest public funds as well.

Dilemmas of technology imports

In the face of so many achievements, the way ISRO handled the acquisition of cryogenic technology stands out as an instance of flawed decision making. The question arises as to why ISRO went against its fundamental ethos and attempted to import cryogenic technology when it had an indigenous development option. I believe the answer lies in the preference for importing technology which developed in the liquid propulsion group after the Viking engine deal with France.

Technology import does not necessarily have to weaken the resolve for indigenous development. The Centaure deal in the Sixties gave ISRO the first production capability in composite solid propellants. But ISRO's own solid propulsion groups were then rapidly able to improve and build on this technology. The energy of the PVC propellants was increased, new higher energy propellant formulations were developed, chemicals for the propellants were produced indigenously, bigger motors were cast, composite motor casings were made, and the technology for segmented motors perfected. The first indigenous Centaures were flown in 1969. Ten years later, all four solid stages of the SLV-3 were ready for flight. Indigenous R&D had achieved a quantum jump over the technology which had been imported. Solids have gone from strength to strength and never looked for outside help again.

It is nearly fifteen years since the first Vikas engine was successfully tested in France. The bottom line is that in this period no similar leap has been achieved in liquids.

After the Centaure import, when the big challenge arose of developing the solid propulsion technology needed to build an all-solid launch vehicle, the issue of importing the technology never even arose. The task took nearly a decade, but it created a strong and self-confident team in solid propulsion. Unfortunately, within the internal dynamics of the liquid propulsion group, the successful acquisition of Viking technology created a strong lobby which had tasted the benefits of technology import. The result was that when the challenge of cryogenic technology loomed ahead, the voices of those who favoured indigenous development were drowned by those who wanted to import the technology. If the gauntlet of indigenous development had been picked up then, ISRO would today have had a very different liquid propulsion group.

How was it that the import of Centaure technology spurred the vigorous growth of indigenous capability in solid propulsion while the import of the Viking technology appears to have had the opposite effect in liquid propulsion? The answer seems to lie in the fact that solids had strong leadership, backed by organizational support, which pushed experimentation and indigenous development over the technology import. Liquids seems to have lacked both.

Although, development of liquid engines started later than solids in ISRO, the early years were full of experiments and promise, starting from primitive pressure-fed engines to flight-tested liquid stages. It is striking that all such experimentation came to a halt after the Viking technology acquisition was completed. Thereafter, the liquid propulsion group has worked only in a project mode: development is undertaken only when a project is formally sanctioned and funds allotted from the ISRO budget.

This does not seem to have been the case with solids. Some level of experimentation to try out ideas and concepts seems to go on, alongside the bread-and-butter project work. Unlike a full-scale project, such experimentation requires very limited funding. What is essential is leadership which can create an environment for innovation.

An example of such experimentation is the building-block technique developed for solid motors. Instead of casting a single solid propellant grain, the propellant slurry is cast into blocks, rather like bricks. The bricks can then be assembled, using more propellant slurry as the mortar. The technique has advantages such as repairing a damaged grain and for building very large solid motors. Through the latter half of the Eighties and into the early Nineties, when the solid propulsion people were busy getting the PSLV's solid first and third stages ready, the annual reports speak of experiments with building-block technology. ISRO has never yet seriously used the building-block technique. But the important point is that innovation was given a chance and in-house capability built up at negligible cost to the organization.

Gowariker had established a Cryogenic Techniques Project as early as 1971, supported the demonstration of a primitive semi-cryogenic engine, and even proposed to Sarabhai the building of a 60 tonne thrust semi-cryogenic engine. Once this cryogenic team was moved to the liquids group in 1974, no further work was done in semi-cryogenics.

Again, although the detailed *Cryogenic Systems Studies* report was submitted in December 1983, basic work on cryogenics, such as building subscale engines to get some experience and derive design data, began only in 1986 when Rs 16.30 crore had been allotted by ISRO as pre-project funding. With leadership which had the drive and the vision, such work could have been taken up much earlier, at the very least immediately after the *Cryogenic Systems Studies* report was submitted. As argued in the previous chapter, this would have allowed ISRO to start full-scale development of an indigenous cryogenic engine by the late Eighties. ISRO could have had its very own cryogenic engine and stage flying by now.

With hindsight, it is also clear that ISRO management

did not adequately appreciate the importance of building on the Viking technology which had been imported. It is not enough to simply import technology and show that the system can be successfully made in India. At various times, proposals were made to improve the Vikas engine itself, including making it regeneratively cooled so that it could operate for much longer durations and also to cluster the engine, as Ariane had done for the first stage of its launcher. But none of these were approved.

The Liquid Propulsion Systems Centre (LPSC) did, for instance, try to persuade ISRO to have a cluster of Vikas engines for the GSLV's first stage. But, after studying over 200 alternative possibilities, the configuration chosen for the GSLV involved having a cryogenic stage replace the upper two stages of the PSLV and four liquid strap-ons in place of PSLV's six solid strap-ons. This decision cannot be faulted as the configuration presented the quickest and lowest cost route to achieving GSLV's goal of putting Insat class satellites into geostationary transfer orbit. A first stage with clustered liquid engines could have hiked the GSLV's project cost by as much as 20 per cent.

These proposals to improve the Vikas engine and cluster them may not have been the quickest or cheapest method to achieve ISRO's immediate objectives. But, in retrospect, it is clear that implementing these measures would have proved beneficial in the longer term, both in terms of economics as well as building up technical capability. Unfortunately, it is only now, in the proposed GSLV Mark-III, that some of these long-standing ideas are likely to be incorporated. It is admittedly not easy for any strongly goal-oriented organization to balance immediate objectives and long-term requirements within its budget. One of the key tasks of the top management must be to make sure that sufficient investments are made in the latter to ensure the continued growth of the organization.

The lesson from the ISRO experience is that importing technology, especially strategic technology, is a two-edged sword. Even when such import appears to have been 'successful', as with the Viking deal, it can tilt the balance of power within an R&D group in favour of further technology imports, rather than take on difficult R&D. Once that happens, tilting it the other way is not so easy. So particular care has to be exercised right from the beginning to ensure that the technology import does not become a crutch, but rather a springboard to further technological growth. Likewise, the measure of success in importing technology cannot be whether an identical system can be manufactured within the country but how much further the technology has thereafter been taken. Strong leadership and good management hold the key to beneficial technology import.

The American embargo which led to the Russians cancelling the technology transfer for cryogenics has had an unintended but beneficial effect. It made quite clear to ISRO's liquid propulsion group that no further technology imports are to be expected. The proposed GSLV Mark-III, which will probably have a clustered liquid first stage and a more powerful cryogenic upper stage, will necessarily have to be entirely based on indigenous efforts.

The road ahead

Anticipating the future and preparing for it have been vital ingredients in ISRO's success. Failure to do this has created the problems which ISRO today faces over cryogenic technology and the adequacy of the GSLV.

The decision to import cryogenic technology was bad enough. When, in 1993, the Russians withdrew from the contract to provide cryogenic technology, ISRO had the opportunity to re-examine the adequacy of the GSLV configuration and the sort of cryogenic stage which it should develop. It opted for a cryogenic stage very similar to the imported Russian one. The GSLV Mark-I (with the Russian cryo stage) does not give the performance originally promised and is now quite inadequate. Worse still, its successor, the Mark-II (with the Indian cryo stage) cannot launch the sort of communication satellites ISRO may itself have to build this decade in order to remain competitive.

It is only recently that an entirely new GSLV configuration with a more powerful cyrogenic stage, the GSLV Mark-III, has been considered. The GSLV Mark-III has not yet been cleared as a project and is unlikely to fly before the end of this decade.

If such mistakes are not to be repeated, ISRO needs a clear overall vision for the future to guide its planning and decision making just as the Sarabhai vision has acted as a lodestone for India's space programme over the last four decades. Globalization, the breakdown of national barriers and growing international cooperation are likely to bring with them new difficulties as well as fresh opportunities in the years ahead. ISRO will need to build on its unique strengths and capabilities so that it can continue to grow in the decades ahead. Many of the principles Sarabhai espoused, such as the emphasis on applications which address major problems facing the country, will undoubtedly continue to be relevant. On the other hand, the sort of self-sufficiency which ISRO has sought to achieve in applications, satellites and launch vehicles may not always be possible. ISRO may need to concentrate its resources, both financial as well as human, in order to achieve international competitiveness in selected areas.

ISRO may have stumbled over the cryogenic issue, but it remains an organization with considerable vitality. It is one of the few institutions in the country with an excellent track record of delivering useful products and services. Over nearly four decades of its existence, it has built up technological strength in applications, satellites and launch vehicles. More important, it has well-developed systems for studying future options as well as for good project management involving multi-disciplinary teams and industry. ISRO needs to bring all these capabilities to bear and create a coherent, carefully thought-out strategy for the future.

The cryogenic experience suggests that ISRO needs to change its approach towards the development and acquisition of technology. The Indian market, whether for applications, satellites or launch vehicles, will be subject to competition from international systems. Such competition will increasingly restrict the margin for error and ISRO will have to make sure that its products and services remain competitive. It will require the ability to initiate technology development sufficiently in advance.

It is worth pointing out that the Agrani satellite to provide mobile telephony in India is being built by a US company and will be launched abroad. Interestingly, two of the senior executives of ASC Enterprises, the company which will operate the Agrani satellite, have headed the Insat Programme Office at ISRO headquarters. But the fact remains that ISRO will not be building or launching the satellite. The Agrani should not be a grim portend of the future.

In launch vehicles, an important issue which will arise

is what sort of launch vehicles are needed and how to build them. The GSLV Mark-III will probably take about ten years to design and built. If it is not internationally competitive, the considerable amount of time and money invested in it may go waste. As it is, the international launcher market is already overcrowded with American, Russian, European and Chinese launchers. Moreover, the United States has already set in motion steps to develop technology which could drastically reduce the cost of access to space, and the Europeans, in order to retain their competitiveness in the commercial launch market, are likely to follow suit.

The principal reason for the current high cost of launch is simply that each launch vehicle, which costs many tens of millions of dollars to build, is used just once. Typically, 2 to 3 per cent of a launcher's lift-off weight can be put into low-earth orbit. The remaining 97 to 98 per cent of the launch vehicle is 'expendable' — it is jettisoned during flight and thrown away. That is like discarding an aircraft, whether it be a small Cessna or the Boeing 747, after its maiden flight! The result is that, as an article in the *Aeronautical Journal* commented, expendable launchers cost 10,000 times more per flight than airliners.

The United States has taken the lead in developing a new breed of reusable launch vehicles (RLVs). These first-generation RLVs would rely on substantial improvements in current technology. They would be designed right from the beginning for minimal maintenance requirements with small ground crews and quick turnaround. The first-generation RLVs, which could be operational by the end of the decade, are expected to cut the current launch cost of $22,000 per kg by a factor of ten.

It is hoped that these first-generation RLVs will pave

the way for true space planes. Using advanced propulsion systems, materials and other technologies, these space planes would operate like the aircraft of today. The space planes could become operational by 2020 and bring the launch cost down to $200 per kg.

Alarmed that even the first-generation RLVs of the United States could undermine Europe's launch capability, the European Space Agency (ESA) launched the Future European Space Transportation Investigations Programme (FESTIP) in late 1994 to study options for a reusable launch vehicle. On the basis of the findings of FESTIP, the agency has just embarked on the first phase of the Future Launchers Technologies Programme (FLTP) to begin development of key technologies. The second phase is likely involve the building of an experimental vehicle to test these technologies in actual flight. Only after that, possibly around 2007, would ESA consider development of a new European reusable launcher. Such a launcher may become operational by 2017-20.

It is important to recognize that operational first-generation RLVs capable of taking on the established launchers like Delta of the United States, Europe's Ariane and Russia's Proton are probably at least ten to fifteen years away. So, while the RLVs may not be an immediate threat, they do pose a long-term challenge which ISRO cannot afford to ignore.

Deciding the sort of GSLV Mark-III and future launch vehicles ISRO should aim for requires much more than looking at other configurations and propulsion systems, reusable or otherwise. Some fundamental issues will have to be examined in detail. One of these will be whether the expense of developing a launcher can continue to be justified in terms of meeting Indian launch requirements alone.

Answering this question will require assessing both the country's own needs as well as the probable growth of the international launch market. In this context, possibilities for working with other countries to jointly develop future launch systems so as to reduce the financial burden, pool technical capabilities and increase marketing reach may also need to be addressed. Within the country too, ISRO may have to team up with various other agencies involved in aerospace development rather attempt to undertake these tasks on its own.

The past four decades have seen the transformation of ISRO, growing from a small band of committed individuals into a large organization with many work centres, programmes, projects and employing nearly 20,000 people. In the early days, the emphasis was on R&D and developing the technology needed to build operational systems. There were dynamic leaders at various levels who were able to understand technology trends and decide the best development path for ISRO. As ISRO grew and its systems entered the operational phase, organizational structures were put in place to ensure the timely delivery of products and services.

Unfortunately, the greater emphasis on projects and project schedules also seems to have weakened systems to look at longer term needs, the technology likely to be required, and the best ways to acquire such technology. Groups have become focused on their particular speciality and power centers have developed within these groups. Charting a course of action for ISRO, which necessarily involves setting priorities and reconciling differences between conflicting views and interests, appears too often to take place only at the chairman's level. As a consequence, there seems to be a preoccupation with influencing the chairman's perceptions rather than evolving a consensus through interactions at various levels between groups.

Despite this, the presence of the old guard, with the knowledge and experience they possess, has played a crucial role in protecting ISRO's interests. Their sense of the larger good of the organization and the camaraderie which still links those of this generation has usually overcome petty jealousies and internal feuds. These people, whose quiet dedication and hard work have made the space programme what it is today, are ageing. Many have already retired and others will do so over the next few years, and the baton will pass to a younger generation.

The coming years are likely to bring complex problems. Planning ahead will require more than just an understanding of technology and technology trends. Charting a course for ISRO will require such knowledge to be integrated with probable market requirements in India and abroad, close coordination with industry and other organizations within the country as well as the ability to handle transnational alliances in the face of geopolitical realities. Institutional mechanisms within ISRO need to be strengthened, and if necessary fresh ones created, to systematically and continually address such multidimensional issues. It will also be essential to maintain a balance between immediate operational requirements and long-term organizational needs. If this can be achieved, it could be one of the most important legacies which the present leaders of the space programme can leave for those who will inherit their mantle.

The decades ahead offer a mixture of uncertainity and unprecedented opportunity. As nations integrate, it is becoming increasingly necessary to address global markets rather than just domestic requirements. In India too, the dogma of self-sufficiency is giving way to realization of the importance of international competitiveness. These trends will inevitably affect space activities too. The challenge for ISRO will be to use the capabilities which have made it a successful national body and transform itself into one to be reckoned with on the international stage.

Notes & Bibliography

THE MOST IMPORTANT resource in writing this book has been access to many past and serving ISRO staff. In the course of interviews and discussions, they explained the complexities and nuances of the technology involved. In ISRO, decisions are taken after considerable debate and study. So talking to the various people who participated in the process provided invaluable insights into how and why many decisions were made. The ISRO specialists were candid about the problems and critical when they thought the organization had erred.

To the extent possible, I have tried to support and supplement the inputs provided by these experts with information from publications of the Department of Space/Indian Space Research Organization. The most important and authentic of these are the Department of Space's *Annual Reports* and *Performance Budgets* which are tabled in Parliament around March-April each year, along with other documents relating to the Union government's budget. The *Annual Report* gives details of the activities and achievements of the Department of Space and ISRO during the previous financial year (April to March). Since the space programme began under the Department of Atomic Energy, documentation of the early years are to be found in the latter's *Annual Reports* and its publication *Nuclear India*.

The *Performance Budget* gives details of each project's sanctioned cost, expenditure on it to date and the projected expenditure during the coming financial year. The month and year when a project was sanctioned can usually be found in the *Performance Budget*. Information about the progress of a project and proposed work on it during the forthcoming financial year are often included. Some of this information may not be available in the *Annual Report*. In addition, inclusion of pre-project funding in the *Performance Budget* indicates activities which ISRO is likely to take up in the form of a full project.

In the document sources for each chapter which follows, I have not included references to the *Annual Reports* and *Performance Budgets*. Inclusion of these would have needlessly lengthened the bibliography. Moreover, where the information provided by them was crucial, reference to the particular financial year to which they pertain has usually been included in the main text itself.

A number of ISRO centres and project groups bring out house journals which are very informative. The ones pertaining to the launch vehicle programme are:

- *Countdown*. This is the house journal of the Vikram Sarabhai Space Centre (VSSC) and was the first of such house journals. The first issue came out in April 1980.
- *Propulsion Today*. The house journal of the liquid propulsion group, now the Liquid Propulsion Systems Centre (LPSC).
- *SHAR News*. The house journal of Shar Centre.
- *PSLV Progress*. The house journal of the PSLV project.
- *GSLV Bulletin*. The house journal of the GSLV project.

In addition, ISRO headquarters has been periodically bringing out *Space India* since 1987, with detailed articles about various activities.

Papers published by ISRO scientists and engineers at conferences and in scientific journals are important sources of technical information. In recent years, papers relating to projects and activities of the space programme have been appearing regularly in *Current Science*, a journal of the Indian Academy of Sciences, Bangalore.

From time to time, ISRO brings out brochures on its various projects and missions which provide much information. These brochures provide a good overview as well as considerable technical data.

The speeches and writings of successive ISRO chairmen give important insights into the thinking of the person holding the top job in the space programme. Some of these which I have used include:

- *Science Policy and National Development*, a compilation of Vikram Sarabhai's papers, edited by Kamla Chowdhry, published by Macmillan India in 1974.
- *Prof. S. Dhawan's Articles, Papers and Lectures (November 1966 to December 1994)*, published by the Indian Space Research Organization, July 1997.
- *Indian Launch Vehicle Development*, the eighth Professor Brahm Prakash Memorial Lecture delivered by Prof. U.R. Rao, The Indian Institute of Metals, Bangalore Chapter, August 1992.
- *Satellite Launch Vehicle Development: An Experience in Self-Reliance*, Faraday Memorial Lecture delivered by Dr K. Kasturirangan in November 1994.

20 Years of Rocketry in Thumba: 1963-1983 published by the Vikram Sarabhai Space Centre in December 1983 gives a comprehensive picture of progress in various areas related to rocketry up to the early Eighties. By then, the SLV-3 had been launched and work had begun on the Augmented Satellite Launch Vehicle (ASLV) and the Polar Satellite Launch Vehicle (PSLV).

There are two important works by non-ISRO people which have looked at the Indian space programme:

- *Technology Development in India's Space Programme 1965-1995: The Impact of the Missile Technology Control Regime*, a 1998 University of Sussex doctoral thesis by Dr A. Baskaran, gives a detailed account of ISRO's technology development. Dr Baskaran has covered satellites as well as launch vehicles and also closely examined industrial involvement in the space programme.

- *Space Today* by Mohan Sundara Rajan, published by the National Book Trust, India (second revised edition, 1995) places Indian efforts in space science, applications, satellites and launch vehicles in their international context. The book also explains the basic principles involved in such development.

Since much of this book deals with rocket propulsion systems, an important reference work was the classic *Rocket Propulsion Elements: An Introduction to the Engineering of Rockets* by George P. Sutton and published by John Wiley & Sons. I used the sixth edition published in 1992.

There are many books dealing with the development and growth of rocketry in the world. One such is *The Rocket: The History and Development of Rocket and Missile Technology* by David Baker and published by New Cavendish Books in 1978.

Chapter 1

The Sarabhai Vision

The man behind the name

Many biographical details and reminiscences about Vikram Sarabhai have been drawn from:

- *Vikram Sarabhai: The Man and the Vision*, edited by Padmanabh K. Joshi and published by Mapin Publishing Pvt. Ltd., Ahmedabad.
- Kamla Chowdhry's introduction in *Science Policy and National Development*, a compilation of Vikram Sarabhai's papers, edited by Kamla Choudhry, published by Macmillan India in 1974.

The paraphrased parts in this section drawn from *Vikram Sarabhai: The Man and the Vision* are:

- 'simple life of refinement and culture' from 'The Boy Vikram' by C.J. Bhatt.
- 'a handsome young boy with a lovable personality, pleasant manner, courteous behaviour and a sharp intelligence', from 'My Student, Employer and Friend' by J.S. Badami.
- 'On their first meeting out', from 'My Father Vikram by Mallika Sarabhai.
- 'It was able to show results in just three years', from 'Institution Builder', by Kamla Chowdhry.

Prof. S. Ramaseshan's recollections of his first meeting with Vikram Sarabhai and of the friendship between Sarabhai and Bhabha are drawn from 'Some Random Thoughts on Vikram Sarabhai, C.V. Raman and Homi Bhabha', a talk given by him at the Vikram Sarabhai Space Centre on 12 August 1996 and reproduced in the *Countdown* issue nos. 195-198, July-October 1996.

Starting a sounding rocket station

I am grateful to Dr E.V. Chitnis for giving me a vivid account of how Thumba was selected as the site for the sounding rocket station.

The Bhabha Memorial Lecture on 'Opportunities for Space Research in India' delivered by Vikram Sarabhai at the Indian Rocket Society symposium on 29 December 1969 was reprinted in the *Journal of the Indian Rocket Society*, Vol. 1, No. 1, January 1971. The tale about Vellana Thuruthu was also taken from this talk.

K. Madhavan Nair, then district collector of Trivandrum, wrote about how land acquisition was carried out at Thumba and Veli in a piece titled 'Land Acquisition at Thumba' which appeared in *20 Years of Rocketry in Thumba: 1963-1983*, published by the VSSC in December 1983.

I have made use of R.D. John's 'Some Reminiscences on Space Construction Programme', *Forerunner*, June 1989.

The first launch from Thumba

R. Aravamudan's recollections ('Coming straight from NASA ...) were taken from his piece 'Then ... And Now' published in *20 Years of Rocketry in Thumba: 1963-1983*, VSSC, December 1983.

Sarabhai wrote about the first sounding rocket launch from Thumba in 'Significance of Sounding Rocket Range in Kerala', *Nuclear India*, December 1963.

The need for a space programme

A.P.J. Abdul Kalam's recollections ('after the successful launch of Nike-Apache') from *Wings of Fire: An Autobiography*, A.P.J. Abdul Kalam with Arun Tiwari, Universities Press, 1999.

Dr Homi Bhabha's inauguration of the International Space Physics seminar at Ahmedabad on 28 January 1963 is covered in *Nuclear India*, February 1963.

Sarabhai's speech at the dedication of the Thumba Equatorial Rocket Launching Station to the United Nations in February 1968 is quoted in many ISRO publications. This quotation can be found, for instance, in the *National Paper of India for The Second United Nations Conference on Exploration and Peaceful Uses of Outer Space (Unispace 82)*, page 31.

'I have a dream, a fantasy maybe', are drawn from Kamla Chowdhury's introduction in *Science Policy and National Development*, a compilation of Vikram Sarabhai's papers, edited by Kamala Chowdhury, published by Macmillan India in 1974.

The book *Science Policy and National Development*, a compilation of Sarabhai's papers edited by Kamla Chowdhury, published by Macmillan India, 1974, brings together many of Sarabhai's speeches and writings. It is a useful source for understanding Sarabhai's thinking. The reference to Sarabhai's papers 'In a paper presented at a conference in 1969, Sarabhai...' draws on material from 'Television for Development', a paper presented at the Society for International Development Conference, New Delhi, November 1969.

The *Atomic Energy and Space Research: A Profile for the Decade 1970-80* was published by the Atomic Energy Commission in 1970.

A number of Sarabhai's writings provide testimony of how he envisaged the space applications which would most benefit India. These papers have been conveniently assembled in the book *Science Policy and National Development*. The relevant chapters are:

a) Chapter Three, 'Space Activity for Developing Countries'.
b) Chapter Four, 'Peaceful Uses of Outer Space'. Address given by Sarabhai as scientific chairman of the United Nations Conference on the Exploration and Peaceful Uses of Outer Space, Vienna, August 1968.
c) Chapter Five, 'Television for Development'. Talk at the Society for International Development Conference, New Delhi, November 1969.
d) Chapter Six, 'INSAT — A National System for Television & Telecommunication'. A joint paper by Vikram Sarabhai, E.V. Chitnis, B.S. Rao, P.P. Kale and K.S. Karnik. Presented by Sarabhai at the National Conference on Electronics at Bombay in March 1970.
e) Chapter Seven, 'Remote Sensing in the Service of Developing Countries'. Presidential Address delivered at the Eighth Annual Meeting of the Indian Geophysical Union in December 1970.

The early days of SSTC

Apart from interviews, I used a number of published reminiscences:

- The VSSC publication *20 Years of Rocketry in Thumba: 1963-*

1983 carries some of them. These include Abdul Kalam's 'Down the Memory Lane', 'Then....And Now' by R. Aravamudan, 'Some Reminiscences' by D. Easwardas, 'My Reminiscences of Early Thumba' by R.D. John, 'My Recollection of Events of 20 Years Ago' by A.S. Prakasa Rao, and 'Land Acquisition at Thumba' by K. Madhavan Nair.

- *Wings of Fire: An Autobiography*, A.P.J. Abdul Kalam with Arun Tiwari, Universities Press, 1999.
- 'How it was in those days', an address given by H.G.S. Murthy to the Scientific Community Forum in November 1988 and reproduced in *Countdown*, No. 104, December 1988.
- 'Some Reminiscences on Space Construction Programme' by R.D. John, *Forerunner*, June 1989.
- 'Sentinel of Our Space History', *Countdown*, No. 6, October 1980. An article about the St Mary Magadelene Church at Thumba.
- 'The Bicycle Era', *Countdown*, No. 2, June 1980.

The account of Sarabhai's way of dealing with people is taken from Mallika Sarabhai's 'My Father Vikram' and P.D. Bhavsar's 'There is Nobody to be Ruled', both of which were published in book *Vikram Sarabhai: The Man and the Vision*.

Chapter 2

First Steps in Rocketry

This period was the most difficult to reconstruct because it is so poorly documented. I have relied heavily on extensive interviews with various people who were involved in the programme at the time. I am grateful to Dr A.C. Bahl, currently head of the Central Documentation Division at VSSC, for giving me information from the personal log he maintained of sounding rocket launches from Thumba.

In addition, the following publications were useful:

- The *Atomic Energy and Space Research: A Profile for the Decade 1970-80*, published by the Atomic Energy Commission in 1970.
- The Department of Atomic Energy's *Annual Reports* from 1962-63 to 1971-72.
- 'ISRO's Solid Rocket Motors', paper presented by R. Nagappa, M.R. Kurup and A.E. Muthunayagam at the 39th Congress of the International Astronautical Federation in October 1988 and subsequently published in *Acta Astronautica*, Vol. 19, No. 8, 1989, pp 681-697.
- *Sounding Rockets from ISRO*, a brochure published by ISRO headquarters.
- 'Rohini Sounding Rocket (RSR) Programme', *Indian Space Programme*, brochure published by ISRO headquarters.
- *Countdown:*
  a) 'VSSC tests paper wound rocket', No. 2, June 1980.
  b) 'Rockets for Monex', No. 3, July 1980.
  c) 'Progress in rocket propulsion at VSSC', No. 9, January 1981.
  d) 'Inflight measurements', No. 34, February 1983.
  e) 'IPP-40: The new composite solid propellant', No. 36, April 1983.
  f) 'Chairman at VSSC', No. 40, August 1983.
  g) 'Successful static test of RH-300 Mk-II motor', No. 63, July 1985.
  h) 'IMAP Campign', No. 78, October 1986.
  i) 'RH-300 Mk-II successful maiden flight', No. 86, June 1987.
  j) 'RH-300 Mk-II second successful flight', No. 89, September 1987.
  k) 'Rohini Sounding Rocket Programme', No. 100, August 1988.
  l) 'RH-560 launches for study of plasma depletion (holes) in equatorial ionosphere', Nos. 153-156, January-April 1993.

In putting together information about the Centaure technology acquisition, I benefited from discussions with the late Dr M.R. Kurup who headed the effort and with Dr M.C. Uttam, currently deputy director at VSSC. The details of the first indigenous Centaures were taken from the Department of Atomic Energy's *Annual Report for 1969-70.*

Chapter 3

SLV-3: India's First Launch Vehicle

- *Annual Reports* of the Department of Atomic Energy up to 1971-72.
- *Atomic Energy and Space Research: A Profile for the Decade 1970-80*, The Atomic Energy Commission, 1970.
- *Space Research in India: 1971-72*, Indian Space Research Organization.
- *Annual Reports* of the Department of Space from 1972-73.
- *Performance Budgets* of the Department of Space from 1975-76.
- 'ISRO's Solid Rocket Motors', paper presented by R. Nagappa, M.R. Kurup and A.E. Muthunayagam at the 39th Congress of the International Astronautical Federation in October 1988 and subsequently published in *Acta Astronautica*, Vol. 19, No. 8, 1989, pp 681-697.
- *20 Years of Rocketry in Thumba: 1963-1983*, published by VSSC, December 1983.
- *India's First Satellite Launch Vehicle, SLV-3: The Story of its Development*, brochure published by VSSC.
- *SLV-3: The First Developmental Flight*, brochure published by VSSC.
- *Indian Space Programme*, brochure published by ISRO headquarters.
- *Wings of Fire: An Autobiography*, A.P.J. Abdul Kalam with Arun Tiwari.

- Foreword in *Developments in Fluid Mechanics and Space Technology*, a festschrift dedicated to Prof. Satish Dhawan, edited by R. Narasimha and A.P.J. Abdul Kalam, published by the Indian Academy of Sciences, 1988. For biographical details about Dr Dhawan.
- *Biographical Memoirs of Fellows of the Indian National Science Academy*, Vol. 16, 1993. For biographical details about Dr Brahm Prakash.
- *Countdown:*

  a) 'Our Centre—a profile', No. 1, April 1980
  b) No. 2, June 1980
  c) 'At Sriharikota...' and 'At Our Centre... Missed Heart Beats and Then Shouts of Joy', No. 3, July 1980.
  d) No. 4, August 1980.
  e) No. 5, September 1980.
  f) 'Behind the scenes', No. 6, October 1980.
  g) 'National honours for ISRO', No. 10, February 1981
  h) 'SLV-3 Developmental Flights', No. 13, May 1981.
  i) 'The story of SLV-3 D-1', No. 14, June 1981.
  j) 'Getting ahead with SLV-3 and ASLV', No. 20, December 1981
  k) 'Shri A.P.J. Abdul Kalam leaves VSSC', No. 25, May 1982
  l) 'SLV-3-D-2—The Grand Finale', No. 37, May 1983
  m) 'Citizens of Trivandrum honour Director, VSSC', No. 46, February 1984
  n) 'SLV-3, India's First Generation Satellite Launch Vehicle', No. 100, August 1988.
  o) Issue on ISRO launch vehicles, Nos. 149-152, September-December 1992.

Information about the Scout launch vehicle of the United States on which the SLV-3 was modelled:

- 'Scout—NASA's Small Satellite Launcher', Andrew Wilson, *Spaceflight*, Vol. 21, 11 November 1979, pp 446-459.
- 'The Scout Launcher... An Update', Andrew Wilson, *Journal of the British Interplanetary Society*, Vol. 34, 1981, pp 193-195.
- 'The Scout Launch Vehicle', Jonathan McDowell, *Journal of The British Interplanetary Society*, Vol. 47, 1994, pp 99-108.
- *The Scout*, LTV Astronautics Division, 1965. I am grateful to Dr Jonathan McDowell of the Harvard-Smithsonian Center for Astrophysics for mailing me a photocopy of this user manual.
- *Jane's Space Directory* for 1994-95 and earlier years, published by the Jane's Information Group, United Kingdom.
- From the internet:

  a) Mark Wade's *Encyclopedia Astronautica*, http://www.friends-partners.org/~mwade/spaceflt.htm

  b) *Fact Sheet on the Scout launch vehicle*, NASA Goddard Space Flight Center, http://pao.gsfc.nasa.gov/gsfc/service/gallery/fact_sheets/general/scout.htm

  c) *Scout Launch Vehicle Program*, NASA's Langley Research Center, http://oea.larc.nasa.gov/PAIS/Scout.html

Sarabhai's speech as scientific chairman of the United Nations Conference on the Exploration and Peaceful Uses of Outer Space in 1968 is included as Chapter Four, 'Peaceful Uses of Outer Space', in *Science Policy and National Development.*

'Some Reminiscences on Space Construction Programme' by R.D. John, chief engineer for civil engineering, *Forerunner*, June 1989, gives a graphic account of Sriharikota and also about Sarabhai's first visit there. In addition, ISRO headquarters has published a brochure, *SHAR: India's Space Port.*

The excerpt where Kalam narrates, 'Suddenly the spell was broken...' is taken from *Wings of Fire: An Autobiography.*

Chapter 4

Developing Competence in Solid Propulsion

- *Actualites*, publication of the Société Européenne de Propulsion (SEP), 4e trimestre 1973. Carries item about ISRO-SEP contract for technical assistance for setting up STEX.
- 'Development of solid propellant technology in India', M.R. Kurup, V.N. Krishnamoorthy and M.C. Uttam, *Developments in Fluid Mechanics and Space Technology*, edited by R. Narasimha and A.P.J. Abdul Kalam, published by the Indian Academy of Sciences, 1988.
- 'ISRO's Solid Rocket Motors', paper presented by R. Nagappa, M.R. Kurup and A.E. Muthunayagam at the 39th Congress of the International Astronautical Federation in October 1988 and subsequently published in *Acta Astronautica*, Vol. 19, No. 8, 1989, pp 681-697.
- 'Know about propellant end trimming machine', *Countdown*, No. 76, August 1986.
- 'Know about propellant grain processing', M.C. Uttam, *Countdown*, No. 119, March 1990.
- *20 Years of Rocketry in Thumba: 1963-1983*, published by VSSC, December 1983.
- *Indian Space Programme*, brochure published by ISRO headquarters.
- 'SLV-3, India's First Generation Satellite Launch Vehicle', *Countdown*, No. 100, August 1988.
- ISRO Launch Vehicles, *Countdown*, Nos. 149-152, September-December 1992.
- *Wings of Fire: An Autobiography*, A.P.J. Abdul Kalam with Arun Tiwari.
- *Rocket Propulsion Elements: An Introduction to the Engineering of Rockets*, George P. Sutton, published by John Wiley & Sons, sixth edition, 1992.

HEF-20:

- 'Characterization Study of a New Binder System (Lactone Terminated Polybutadiene—LTPB) for High Energy Solid Propellants', AIAA Paper No. 75-1335, AIAA-SAE 11th Propulsion Conference, 1975.

ISRO polyol:

- 'Low Oxygen Containing Polyesters in Low Cost Polyurethane Propellants for Large Booster Applications', S.K. Nema, A.U. Francis, P.R. Nair and V.R. Gowariker, AIAA Paper No. 76-634, AAIA-SAE 12th Propulsion Conference, 1976.
- UK patent No. 1524782 granted in January 1977 for 'Production of Ester-type Polyols'.
- The United States of America patent No. 4,161,482 granted in July 1979 for 'Production of Polyols Containing Basic Nitrogen'.
- 'ISRO Polyol—The Versatile Binder for Composite Solid Propellants for Launch Vehicles and Missiles', V.N. Krishnamurthy and Solomon Thomas, Journal of Defence Science, Vol. 37, No. 1, pp 29-37, January 1987.
- 'IPP-40: The new composite solid propellant', *Countdown,* No. 36, April 1983.
- 'RH-300 Mk-II successful maiden flight', *Countdown,* no. 86, June 1987.

Space crude:

- 'Space Technology and the Utilization of Forest Wastes', Satish Dhawan, Dr. A.L. Mudaliar Memorial Second Lecture at the Indian Institute of Technology, Madras, on 2 February 1976, reproduced in *Prof. S. Dhawan's Articles, Papers and Lectures (November 1966 to December 1994)*, published by the Indian Space Research Organization, July 1997.

Ammonium perchlorate:

- 'APEP Completes 10 Years', *Countdown*, No. 106, February 1989.

- 'APEP logs 1000 tonnes in production', *Countdown*, No. 123, July 1990

Ariane Passenger Payload Experiment (APPLE):

- *APPLE: Experimental Geostationary Communication Satellite*, brochure published by ISRO.
- *APPLE, Ariane Passenger Payload Experiment: India's First Experimental Communications Satellite*, brochure published by ISRO.
- 'VSSC and Apple', *Countdown*, No. 2, June 1980.
- 'Progress in rocket propulsion at VSSC', *Countdown*, No. 9, January 1981.
- 'APPLE aloft' and 'The role of VSSC', *Countdown*, No. 15, July 1981.
- APPLE put to Nation's Service, *Countdown*, No. 17, September 1981.
- 'Two Years of APPLE', *Countdown*, No. 39, July 1983.

Composites:

- 'Composites development in VSSC', *Countdown*, No. 10, February 1981.
- 'R&D in propellants, chemicals & materials group', *Countdown*, No. 12, April 1981.
- 'It is special phenolic resin, this time', *Countdown*, No. 24, April 1982.

Chapter 5

Early Initiatives in Liquid Propulsion

- 'Development of Liquid Propulsion Systems in ISRO', A.E. Muthunayagam, paper No. IAF-88-224 presented at the 39th International Astronautical Federation, 1988.
- *20 Years of Rocketry in Thumba: 1963-1983*, published by VSSC, December 1983.

- *Rocket Propulsion Elements: An Introduction to the Engineering of Rockets*, George P. Sutton, sixth edition, 1992.

Europa and early Viking development:

- 'The Ariane family story and beyond', *Reaching for the Skies*, ESA BR-42, June 1988, European Space Agency.
- *Revue aerospatiale*, special issue 1990.
- *Ariane: A European Success Story*, European Space Agency, May 1992.
- *Le bulletin: Informations Aéronautiques et Spatiales*, 7 May 1992, No. 1554, Groupement des Industries Françaises Aéronautiques et Spatiales (GIFAS).
- *Info: Ariane 100th Launch*, European Space Agency, August 1997.
- 'From L3S to Arianespace', *Interavia*, December 1999.

Chapter 6:

The ASLV: A Technological Bridge

- 'Augmented Satellite Launch Vehicle', M.B. Reddy and M.S.R. Dev, *Current Science*, Vol. 66, No. 6, pp 408-416, 25 March 1994.
- *ASLV: ISRO's Second Generation Launch Vehicle*, brochure published by the ASLV project, VSSC.
- *ASLV-SROSS Mission*, brochure published by VSSC, November 1986
- *20 Years of Rocketry in Thumba: 1963-1983*, published by VSSC, December 1983.
- 'ASLV: Getting Ready for Lift-Off', *Space India*, January 1987.
- 'ASLV—The Augmented Satellite Launch Vehicle', *Countdown*, No. 100, August 1988.

- 'A glimpse of the future', *Countdown*, No. 7, November 1980.
- 'Getting ahead with SLV-3 and ASLV', *Countdown*, No. 20, December 1981.
- 'Chairman's Reviews', *Countdown*, No. 26, June 1982.
- 'The President is pleased...', *Countdown*, No. 27, July 1982.
- 'AS-4 static test successful', *Countdown*, No. 30, October 1982.
- 'ASLV strap-on jettisoning mechanism successfully tested', *Countdown*, No. 50, June 1984.
- 'ASLV-D1 launch campaign on', *Countdown*, No. 53, September 1984.
- 'Inhouse fabrication of ASLV structures', *Countdown*, No. 57, January 1985.
- 'Focus on ASLV aerodynamics and structures', *Countdown*, No. 60, April 1985.
- 'AS-4 motor tested in HAT facility', *Countdown*, No. 62, June 1985.
- 'Focus on inhouse fabrication for ASLV', *Countdown*, No. 64, August 1985.
- 'Successful strap-on test flight', *Countdown*, No. 68, December 1985.
- 'ASLV Stage-III motor qualified', *Countdown*, No. 68, December 1985.
- Focus on ASLV, *Countdown*, No. 72, April 1986.
- 'ASLV—Heat shield tested in KHS', *Countdown*, No. 75, July 1986.
- 'Prof. U.R.Rao talks to *Countdown*', *Countdown*, No. 96, April 1988.
- 'ASLV strap-on motor successfully tested', *Countdown*, No. 123, July 1990.
- 'AS0 Segments Triple Casting at SPROB', *SHAR News*, No. 29, April-June 1991.
- 'Contol System Packages Integrated with ASLV-D4', *Propulsion Today*, Vol. VI, No. 1, January-March 1994.

First flight:
- 'Focus on SROSS-1', *Countdown*, No. 82, February 1987.
- 'ASLV: Getting Ready for Lift-Off', *Space India*, No. 1, 1987.
- ASLV-D1 Flight, *Countdown*, No. 83, March 1987.
- 'Chairman at VSSC', *Countdown*, No. 84, April 1987.
- 'Failure Analysis of the ASLV-D1 Flight', *Countdown*, No. 87, July 1987.

Second flight:
- ASLV-D2, *Countdown*, No. 99, July 1988.
- Dr. S.C. Gupta Talks to *Countdown* on ASLV-D2 Failure, *Countdown*, No. 110, June 1989.
- *Note on ASLV-D2*, press conference by ISRO chairman, Dr U.R. Rao, at VSSC on 23 August 1988.
- *Press Release* on 'Scientific Departments' Consultative Committee Meets', Press Information Bureau, 10 August 1989.
- *Note on Augmented Satellite Launch Vehicle*, Meeting of the Consultative Committee of Parliament for Scientific Departments, 10 August 1989.
- *Press Release* on ASLV-D3 launch, VSSC, 20 May 1992.

Third flight:
- 'ASLV Launch Campaign Commenced', *SHAR News*, No. 31, October-December 1991.
- 'ASLV-D3 successfully launches SROSS-C into orbit', *Countdown*, No. 145, May 1992.
- 'ASLV Launch Successful', *Space India*, April-June 1992.

Fourth and final flight:
- *ASLV-D4/SROSS-C2*, pamphlet published by ISRO headquarters.
- ASLV-D4, *SHAR News*, No. 40, April-September, 1994.
- 'ASLV-D4 Launch Successful', *Space India*, April-June 1994.
- ASLV-D4, *Countdown*, No. 170, June 1994.

Chapter 7

Guiding a Launcher From Ground to Orbit

- *Inventing Accuracy: A Historical Sociology of Nuclear Missile Guidance*, Donald MacKenzie, The MIT Press, 1993.
- 'Guidance and Control Systems for Satellite Launch Vehicles', R.M. Vasagam and S.C. Gupta, *Journal of the Indian Rocket Society*, Vol. 1, No. 1, January 1971, pp 39-49.
- 'Management of Advanced Technology Development for the Control and Guidance of Launch Vehicles', S.C. Gupta, Hari Om Ashram Prerit Vikram Sarabhai Award Lecture, August 1980.
- 'Development of navigation guidance and control technology for Indian launch vehicles', S.C. Gupta and B.N. Suresh, Sadhana, Vol. 12, Part 3, March 1988, pp 235-249 and reproduced in *Developments in Fluid Mechanics and Space Technology*, edited by R. Narasimha and A.P.J. Abdul Kalam.
- 'Growth of capabilities of India's launch vehicles', S.C. Gupta, *Current Science*, Vol. 68, No. 7, 10 April 1995, pp 687-691.
- 'Overview of Launch Vehicle Flight Control System', B.N. Suresh, Workshop on Flight Control Systems, Bangalore, December 1995.
- 'Fault Tolerant Inertial Guidance System for Indian Satellite Launch Vehicles', B.N. Suresh, paper IAF-98-A.3.01 presented at 49th International Astronautical Congress, 1998.
- *Improved Guidance Hardware Study for the Scout Launch Vehicle*, Roger T. Schappell, Michael L. Salis, Ray Mueller, Lloyd E. Best, Albert J. Bradt, Roger Harrison and John H. Burrell, NASA Contractor Report (NASA CR-2029), June 1972.
- *Beryllium Facilities*, brochure published by the Bhabha Atomic Research Centre.
- *Redundant Attitude Reference Systems (REARS)*, brochure published by ISRO headquarters.

- *Countdown:*

    a) 'SLV-3 E02 trajectory' and 'Technology Transfer from VSSC', No. 5, September 1980.
    b) 'A Glimpse of the Future', No. 7, November 1980.
    c) 'Our efforts in Avionics', No. 8, December 1980
    d) 'Inertial Navigation Systems', No. 39, July 1983.
    e) 'Focus on ASLV Avionics', No. 51, July 1984.
    f) 'ASLV-D1 Launch Campaign On', No. 53, September 1984.
    g) 'Computers in Space' and 'Know about BMF', No. 58, February 1985.
    h) 'ASLV onboard software successfully tested', No. 62, June 1985.
    i) 'Closed loop guidance for ASLV-D1 qualified', No. 74, June 1986.
    j) 'Beryllium Maching Facility Bags Award', No. 120, April 1990.
    k) 'ASLV-D4/SROSS-C2 mission', No. 170, June 1994.
    l) 'Critical Beryllium Components Successfully Produced', No. 177-181, January-May 1995.

- ASLV-D4, *SHAR News*, no. 40, April-September 1994.

Chapter 8

PSLV: Achieving Operational Launch Capability

Dr R.M. Vasagam, who headed the SLV-SYN Study Team which was set up in 1972, gave me details about the launch vehicle configuration which they had recommended. I am grateful to the present ISRO chairman, Dr K. Kasturirangan, for giving me access to some of the early studies and reports which led to the present configuration for the PSLV. The internal documents which I was able to go through were:

- *Report of the Committee on System Studies of Vehicle Configurations Options for SLV Variants*, April 1978. The Committee was headed by Dr S. Srinivasan, later project director for the PSLV. This report also carries the Office Memorandum dated 10 January 1978 issued by Dr Brahm Prakash which laid down the guidelines to be followed for arriving at vehicle configurations.
- *Polar Satellite Launch Vehicle Development Studies*, December 1978, VSSC-PSLV-MR-01-78, prepared by PSLV Study Team.
- *Polar Satellite Launch Vehicle Project*, December 1981, PSLV-VSSC-PJ-PM-04-81, prepared by Dr S. Srinivasan, project director, PSLV.

In addition, the following articles which appeared in *Countdown* also provided insights into the evolution of the PSLV configuration:

- 'A Glimpse of the Future', No. 7, November 1980.
- 'PSLV Takes Shape', No. 19, November 1981.
- 'LAM for PSLV', No. 38, June 1983.
- 'PSLV status review by Chairman', No. 46, February 1984.

The following works give considerable information about the PSLV:

- 'Indian Launch Vehicle Programme', D. Narayana Moorthi. To be published shortly.
- 'Flight Experience of Indian Polar Satellite Launch Vehicle', S. Ramakrishnan and S. Srinivasan, paper No. IAF-98-V.1.05 at 49th International Astronautical Congress, 1998.
- *PSLV: Polar Satellite Launch Vehicle*, brochure published by ISRO headquarters, September 1998.

Solid propulsion

- 'ISRO's Solid Rocket Motors', paper presented by R. Nagappa, M.R. Kurup and A.E. Muthunayagam at the 39th Congress of the International Astronautical Federation in October 1988 and subsequently published in *Acta Astronautica*, Vol. 19, No. 8, 1989, pp 681-697.

The first stage of PSLV:

- 'PSLV—PS1 middle segment, successful proof pressure test', *Countdown*, No. 94, February 1988.
- 'PSLV First Stage Motor (PS-1) Middle Segment Cast at SPROB', *PSLV Progress*, No. 1, October 1988.
- 'PSLV project posts substantial progress', *Countdown*, 103, November 1988.
- 'Processing of Solid Propellant Grain for PS-1 Motor', *PSLV Progress*, No. 2, April 1989.
- 'PSLV First Stage Successfully Tested', *Space India*, April-December 1989,
- 'PSLV Project Crosses a Milestone, First Stage Motor Successfully Tested', *Countdown*, No. 115, November 1989.
- 'PSLV First Stage Motor—Another Successful Test', *Space India*, January-March 1991.
- 'PS-1 static test—a repeat performance', *Countdown*, No. 132, April 1991.
- 'Second Test of PSLV First Stage Motor', *PSLV Progress*, No. 7, June 1991.
- 'First Stage of PSLV Qualified', *SHAR News*, No. 28, January-March 1991.
- 'S-139 Booster Test Successful', *PSLV Progress*, No. 18, May 1997.
- 'S-139 Booster Successfully Static Tested', *SHAR News*, No. 47, November 1996-April 1997.
- 'Improved Rocket Motors Tested', *Space India*, April-June 1997.

The solid third stage:

- 'PSLV status review—SR13', *Countdown*, No. 93, January 1988.
- 'Successful Pressure Test of PS3 Prototype Motorcase', *PSLV Progress*, No. 1, October 1988.
- 'PSLV Third Stage Motor (PS-3) Cast at SPROB', *PSLV Progress*, No. 2, April 1989.

- 'Successful Testing of the PSLV Third Stage Motor', *Countdown*, No. 108, April 1989.
- 'Second Static Test of the PSLV Third Stage (PS-3) Motor', *PSLV Progress*, No. 5, January 1990.
- 'PS3-03 Failed During Static Test', *SHAR News*, No. 24, April-June 1990.
- 'PS3 Motor Fails in Static Test', *SHAR News*, No. 26, October-December 1990.
- 'PS3 Motor Case with Modified Submerged Nozzle Sub-Assembly Pressure Tested', *SHAR News*, No. 29, April-June 1991.
- 'PSLV Third Stage Motor', *Space India*, October 1991-March 1992.
- 'Successful PS3(06) Test Clears the Way for PSLV Flight Preparations', *PSLV Progress*, No. 9, January 1992.
- 'PSLV Third Stage Static Tests (07) & (08) Completed', *PSLV Progress*, No. 11, September 1992.
- 'Ninth and Final Static Test of PS3 Motor Completed', *PSLV Progress*, No. 12, January 1993.
- 'Rubber Products for Space', *Space India*, April-June 1994.

Growth in Solid Propulsion–Graph I

The delta-V has been computed for each stage, based on its mass ratio and its specific impulse in vacuum. The data for many of the ISRO stages were taken from 'ISRO's Solid Rocket Motors', the paper by R. Nagappa, M.R. Kurup and A.E. Muthunayagam. ISRO sources provided the relevant information about the other ISRO solid stages. The data on the Orbus 21D, which forms the third stage of the Athena II rocket, was got from *Mission Overview: Athena II–Ikonos* distributed by Space Imaging and Lockheed Martin at the time of the Ikonos launch in September 1999. Data on the Ariane-5's solid propellant boosters was taken from 'Solid Propellant-Stage Development for Ariane-5', J. Gigou, *ESA Bulletin*, No. 69. The data on the Orbus 21 (the first stage of the Inertial Upper Stage, IUS) and the Space

Shuttle's Solid Rocket Boosters came from *Jane's Space Directory, 1994-95*, published by the Jane's Information Group, UK, 1994.

Liquid propulsion and PSLV's second stage:

- *Liquid Propulsion Systems Centre*, brochure published by LPSC.
- *Second Stage of PSLV—A High Thrust Liquid Propulsion System*, brochure published by LPSC, November 1989.
- 'Vikas engine fabricated', *Countdown*, No. 35, March 1983.
- 'Reflections 1985', *Countdown*, No. 69, January 1986.
- 'Full Scale Integration Mock-Ups of PS-2 & PS-4 Realised', *Propulsion Today*, No. 5, November-December 1986.
- 'Pressure Transducer Fabrication Facility', 'Vikas Engine Contoured Nozzle Thrust Chamber Realised' and 'Titanium Alloy Gas Bottle for PS2 Pressurization system Qualified', *Propulsion Today*, No. 6, January-April 1987.
- 'A Test Stand for Liquid Engines', *Space India*, July-September 1987.
- 'PS2 Stage Tank Realised', 'PS2 Engine Test' and 'PS2 Water Tank Qualified', *Propulsion Today*, No. 7, March 1988.
- 'Pogo System Studies at ATS', *Propulsion Today*, Vol. II, No. 2-3, July-December 1998.
- 'PS-2 Full Duration Test', *PSLV Progress*, No. 2, April 1989.
- 'Proof Pressure Testing of the PSLV Second Stage (PS-2) Propellant Tank', *PSLV Progress*, No. 4, October 1989.
- 'Indigenisation of PS-2 Gas Bottles', *PSLV Progress*, No. 3, July 1989.
- 'Pogo System Tests at ATS', *Propulsion Today*, 1989.
- 'PSLV Second Stage Successfully Tested', *Space India*, January-March 1990.
- 'PSLV Second Stage (PS-2) Battleship Tests Completed' and 'Structural Test on the Second Stage (PS-2) Thrust Frame', *PSLV Progress*, No. 6, August 1990.
- 'Flight Version Stage Qualified', *Propulsion Today*, Vol. IV, No. 3, July-September 1992.

- 'PSLV-D1: PS2 stage', M.K.G. Nair, *Propulsion Today*, vol. V, 1993.
- 'PSLV Second Stage Flight Configuration Qualified', *PSLV Progress*, No. 12, January 1993.
- 'Indigenous Sephen Throat for Vikas Engine Qualified', *Propulsion Today*, Vol. VI, No. 2, April-June 1994.
- 'Indigenous Sephen Throat Qualified for L-40 Stage', *Propulsion Today*, Vol. VII, No. 2, May-August 1995.
- 'Silica-phenolic throat inserts', *Countdown*, Nos. 182-188, June-December 1995.
- 'Throat Insert for Liquid Rocket Engines', *Space India*, July-December 1995.
- 'Stretched Version of PS2 Propellant Tank Delivered', *PSLV Progress*, No. 18, May 1997.
- 'Liquid Propulsion Strap-on Stage of GSLV qualified', press release no. PPR:D:185:98 issued on 5 March 1998 by ISRO headquarters.

PSLV's fourth stage:

- *Liquid Propulsion Systems Centre*, brochure published by LPSC.
- *PS-4*, brochure published by LPSC.
- 'LAM for PSLV', *Countdown*, No. 38, June 1983.
- 'Full Scale Integration Mock-Ups of PS-2 & PS-4 Realised', *Propulsion Today*, No. 5, November-December 1986.
- 'PSLV status review—SR13', *Countdown*, No. 93, January 1988.
- 'PS-4 Engine Tested Successfully', 'PS-4 Ablative Engine Test' and 'Titanium Gas Bottle for PS-4', *Propulsion Today*, Vol. II, April-June 1988.
- 'Major Breakthrough in LUS Engine Development', *PSLV Progress*, No. 1, October 1988.
- 'PSLV Fourth (PS-4) Stage Battleship Version Successfully Tested', *PSLV Progress*, No. 4, October 1989.
- 'Highlights of Progress', *PSLV Progress*, No. 3, July 1989.

- 'PS-4 Battleship Assembly and Hot Test', *Propulsion Today*, 1989.
- 'Fourth Stage (PS-4) Tank Forgings Indigenised', *PSLV Progress*, No. 6, August 1990.
- 'Focus on Fourth Stage of PSLV', *Countdown*, No. 133, May 1991.
- 'PSLV Fourth Stage (PS-4) in Flight Configuration Static Tested', *PSLV Progress*, No. 7, June 1991.
- 'PSLV Fourth Stage Liquid Propellant Motor', *Space India*, April-June 1991.
- 'PS-4 HAT Test Facility Commissioned', *PSLV Progress*, No. 8, September 1991.
- 'PS-4 and Equipment Bay Sub-Assembly Ready for Vibration Test', *PSLV Progress*, No. 10, April 1992.
- 'PS-4 Engine tested in HATF', *Propulsion Today*, Vol. IV, No. 2, April-June 1992.
- 'High Altitude Qualification Testing of PSLV Fourth Stage Engine Completed', *PSLV Progress*, No. 11, September 1992.
- 'Successful ground qualification in flight configuration of the liquid fourth stage of PSLV', press release no. PPR:D:24:92 dated July 27, 1992 issued by ISRO headquarters.
- 'Fourth Stage of PSLV Qualified for Flight', *Propulsion Today*, Vol. IV, No. 3, July-September 1992.
- 'Fourth Stage (PS4) of PSLV', Mohammed Muslim, *Propulsion Today*, Vol. V, 1993.
- 'Fourth Stage Handed Over for PSLV-D1 Launch' and 'First Stage Reaction Control System Qualified', *PSLV Progress*, No. 12, January 1993.
- 'PSLV-C1 Post Flight Analysis Concluded', *PSLV Progress*, No. 20, July 1998.

Inertial guidance

- 'Growth of capabilities of India's launch vehicles', S.C. Gupta, *Current Science*, Vol. 68, No. 7, 10 April 1995, pp 687-691.

- 'Fault Tolerant Inertial Guidance System for Indian Satellite Launch Vehicles', B.N. Suresh, paper IAF-98-A.3.01 presented at 49th International Astronautical Congress, 1998.
- 'RESINS testing in progress', *PSLV Progress,* No. 1, October 1988.
- 'DAP (Digital Autopilot) Design Reviewed', *PSLV Progress,* No. 9, January 1992.
- 'PSLV Control Electronics Packages Developed', *PSLV Progress,* No. 12, January 1993.
- 'PSLV Failure Analysis Committee Submits Report', press release no. PPR:D:72:93 dated 3 January 1994 issued by ISRO headquarters.
- 'From PSLV-D1 to PSLV-D2', *PSLV Progress,* No. 14, October 1994.
- 'Post Flight Analysis Completed: PSLV-D3 Performance Close to Prediction', *PSLV Progress,* No. 17, December 1996.

Industry involvement in PSLV

- *Space-Industry News: Newsletter on the Indian Space Programme's Partnership with Industry*, issued by the Directorate of Technology Transfer & Industry Cooperation at ISRO headquarters, Nos. 1-9, September 1989 to November 1997.
- *Induspace '91/92*, published by the Directorate of Technology Transfer & Industry Cooperation at ISRO headquarters, April 1992.
- *Induspace 1992/93*, published by the Directorate of Technology Transfer & Industry Cooperation at ISRO headquarters, December 1993.
- 'ISRO-BEL collaboration: focus on VSSC's role', *Countdown*, No. 77, September 1986.
- 'Towards self-reliance in hi-rel electronic components', *Countdown*, No. 131, March 1991.
- 'ISRO-HAL Cooperation in Space', *Space India*, July-September 1991.

- 'Focus on ISRO-HAL cooperation', *Countdown*, No. 136, August 1991.
- 'PSLV First Developmental Flight—Preparations in Full Swing', *Space India*, October 1992–March 1993.

Maraging steel:

I benefited from discussions with Dr P. Rama Rao, former director of the Defence Metallurgical Research Laboratory (DMRL) and currently vice-chancellor of the Central University at Hyderabad.

- 'R&D in propellants, chemicals & materials group', *Countdown*, No. 12, April 1981.
- 'PSLV takes shape', *Countdown*, No. 19, November 1981.
- 'Chairman reviews PSLV status', *Countdown*, No. 70, February 1986.
- 'PSLV projects posts significant progress', *Countdown*, No. 81, January 1987.
- 'Ultrasonic testing of maraging steel motor case', *Countdown*, No. 82, February 1987.
- 'PSLV Status Review SR-12', *Countdown*, No. 88, August 1987.
- 'Indigenous Development of 18 Ni 2400 Maraging Steel', V.K. Gupta, M. Chatterjee and R.P. Bhatt, Trans. *Indian Inst. Met.*, Vol. 43, No. 3, June 1990, pp 129-137.
- 'New Heat Treatment for Tougher Maraging Steels', V.K. Gupta and M. Chatterjee, *Advanced Materials & Processes*, Vol. 138, No. 9, 1990, page 90.
- 'PSLV projects posts significant progress', *Countdown*, No. 81, January 1987.
- 'Ultrasonic testing of maraging steel motor case', *Countdown*, No. 82, February 1987.
- 'PSLV Booster: Maraging Steel Hardware Starts Arriving', *Countdown*, No. 91, November 1987.
- 'Development of new low nickel, cobalt free maraging steel', K.T. Tharian, D. Sivakumar, R. Ganesan, P. Balakrishnan

and P.P. Sinha, *Materials Science and Technology*, Vol. 7, December 1991, pp 1082-1087.
- 'Development of Heat Treatment Parameters to Improve Fracture Toughness and Grain Size of an Embrittled Maraging Steel', P.P. Sinha, K. Sreekumar, N.S. Babu, B. Pant, A. Natarajan and K.V. Nagarajan, *J. Heat Treating*, Vol. 9, No. 2, 1992, pp 125-131.
- 'Maraging Steel: Lab to Launch—A Recollection', P.P. Sinha, *Met News*, Vol. 12, No. 1, October 1993.
- 'Indigenous Development of Maraging Steels for Space Applications', Thomas Tharian, D. Sivakumar, K. Sree Kumar, P.P. Sinha and K.V. Nagarajan.
- 'Indigenous Development of High Strength Low Alloy Steels for Space Applications', M.R. Suresh, R. Suresh Kumar, S.C. Sharma and P.P. Sinha, *Proceedings of National Seminar on Emerging Technologies and Indigenisation in the Aerospace Industry*, December 1995.
- *Metals and Alloys from Midhani*, brochure published by the Mishra Dhatu Nigam Limited (Midhani).

HTPB:

- 'Development of HTPB Propellant System for ISRO's Motors', R. Nagappa and M.R. Kurup, paper No. AIAA 90-2331 presented at AIAA/SAE/ASME/ASEE 26th Joint Propulsion Conference, 1990.
- *Proceedings of the Colloquium on HTPB*, edited and compiled by K.S. Sastri, S. Alwan, A. Venugopal and C.R. Dhaveji, published by ISRO, ISRO-VSSC-SP-64-92, 1992.
- 'Subscale Test for HTPB and SITVC', *PSLV Progress*, No. 1, October 1988.

Ground facilities and assembly

- *SHAR, India's Space Port*, brochure published by SHAR Centre.
- *PCMC Radar*, brochure published by ISRO.

- 'Precision Radar for Launch Vehicles', *Space India*, January–March 1990.
- 'The Moving Tower of SHAR—Mobile Service Structure for PSLV', *Space India*, July–September 1990.
- 'Mobile service structure for PSLV', *Countdown*, No. 128, December 1990.
- 'Checkout System for PSLV', *SHAR News*, No. 29, April–June 1991.
- 'Real Time Computer Support for Launch Operations', *SHAR News*, No. 30, July–September 1991.
- 'Precision Radars for PSLV', *SHAR News*, No. 31, October–December 1991.
- 'PCMC Radar: First Radar Commissioned', *PSLV Progress*, No. 9, January 1992.
- 'Brisk Activity at Launch Complex', *SHAR News*, No. 37, April-July 1993.
- 'PSLV Launch—Focus Shifts to SHAR Launch Complex', *Space India*, April–June 1993.

The PSLV gets off the ground

PSLV-D1:

- *PSLV*, brochure published by ISRO headquarters.
- 'PSLV Launch—Focus Shifts to SHAR Launch Complex', *Space India*, April–June 1993.
- 'PSLV First Developmental Flight—Preparations in Full Swing', *Space India*, October 1992–March 1993.
- 'PSLV-D1 mission', U.R. Rao, S.C. Gupta, G. Madhavan Nair and D. Narayana Moorthi, *Current Science*, Vol. 65, No. 7, 10 October 1993, pp 522-528.
- 'First Developmental Flight of PSLV', *Space India*, July–September 1993.
- 'PSLV-D1 Mission', *Propulsion Today*, Vol. V, 1993.
- 'PSLV Failure Analysis Committee Submits Report', press release No. PPR:D:72:93 dated 3 January 1994 issued by ISRO headquarters.

- 'PSLV Failure Analysis Committee Submits Report', *Space India*, October 1993–March 1994.
- 'PSLV-D1 Missions: Findings & Recommendations of FAC', *SHAR News*, No. 39, October 1993–April 1994.

PSLV-D2:

- 'PSLV-D2 Ready for Launch', *PSLV Progress*, No. 14, October 1994.
- *PSLV-D2*, brochure published by ISRO headquarters.
- 'PSLV-D2 Launch Successful', press release no. PPR:D:87:94 dated 15 October 1994 issued by ISRO headquarters.
- 'India's first Polar Satellite IRS-P2 launched by PSLV-D2', *Current Science*, Vol. 67, No. 8, October, 1994, pp 565-570.
- 'Glimpse at PSLV-D2', *Propulsion Today*, Vol. VI, No. 4, October–December 1994.
- 'PSLV-D2 Launch Successful', *Space India*, October–December 1994.

PSLV-D3:

- 'PSLV-D3 Launch Successful', press release no. PPR:D:132:96 dated 21 March 1996 issued by ISRO headquarters.
- 'PSLV-D3 Launch Successful', *Space India*, January–June 1996.
- 'Post Flight Analysis Completed: PSLV-D3 Performance Close to Prediction', *PSLV Progress*, No. 17, December 1996.

PSLV-C1:

- *PSLV-C1/IRS-1D*, brochure issued by ISRO headquarters.
- 'Configuration of PSLV-C1', *PSLV Progress*, No. 17, December 1996.
- 'PSLV Launches IRS-1D', *Space India*, July–September 1997.
- 'PSLV-C1/IRS-1D Mission', *Current Science*, Vol. 73, No. 8, October 25, 1997.
- 'PSLV-C1 Post Flight Analysis Concluded', *PSLV Project*, No. 20, July 1998.

PSLV-C2:

- *PSLV-C2 Mission*, brochure published by PSLV project, VSSC.
- *PSLV-C2*, brochure published by ISRO headquarters.
- 'PSLV-C2 Mission', background material for the Press prepared by ISRO headquarters.
- 'PSLV Successfully Launches Three Satellites', press release no. PPR:D:218:99 dated 26 May 1999 issued by ISRO headquarters.
- 'Passenger Payloads on PSLV-C2: The First Commercial Launch Service by Antrix/ISRO', S. Ramakrishnan, K. Ramachandran and S.S. Balakrishnan, paper no. IAF-99-V.5.08 presented at 50th International Astronautical Congress, October 1999.
- 'PSLV-C2 Mission', *Current Science*, Vol. 77, No. 8, October 25, 1999, pp 1038-1045.
- 'PSLV Enters World Launch Market', S. Ramakrishnan, *PSLV Progress*, No. 22, August 1999.

Chapter 9

The Saga of the Cryogenic Engine

The pre-project funding for cryogenic development appears under the head 'Cryogenics (Engine & Stage)' in the *Performance Budget for 1987-88*, p 53. It is stated there:

> The development of the cryogenic engine is vitally important for attaining geostationary launch capability with the GSLV for launching Insat-II class satellites. Experience elsewhere in the world has shown that cryogenic engine development takes about fifteen yeares. Efforts are being made in ISRO to compress the time schedule to a much lesser period and to realise the cryogenic engine to be available by 1993-94 for the GSLV. Studies have been conducted by ISRO and full project report at a cost of Rs 240 crores for the development of a 12 T Cryogenic Engine and Stage is under critical evaluation. Pending scrutiny and approval of the Project Report, keeping in view

the need to meet the 1993-94 schedule of GSLV, certain minimum investments are required to be initiated now for setting up a Cryogenic Engineering Laboratory for sub-scale cryo engine fabrication and testing so that the inputs may be available for the main 12 T engine development, once the project is approved. A pre-project investment of Rs 16.30 crores has also been approved recently. Of this, a sum of Rs 11.47 crores is for setting up the laboratories and facilities for development, fabrication and testing of the sub-scale engine.

The information about the Chinese development of their YF-73 cryogenic engine was from *World Guide to Commercial Launch Vehicles*, Frank Sietzen, published by Pasha Publications, USA, 1991 and *Jane's Space Directory, 1993-94*, Jane's Information Group, UK.

The *Performance Budget for 1988-89* shows a pre-project allocation of Rs 1 crore for the GSLV during that year. The *Performance Budget* for the following year, 1989-90, indicates an allocation of Rs 4.5 crore for that year and the *Performance Budget for 1990-91* an allocation of Rs 30 crore. It is only in the *Performance Budget for 1991-92* that the sanctioning of the GSLV project at a total cost of Rs 756.07 crore is announced (pages 19-21).

The details of ISRO's deal with Glavkosmos for cryogenic technology were given in a statement made by the Indian Prime Minister in the Rajya Sabha on 18 August 1993. This statement of the Prime Minister was reproduced in *Current Science*, Vol. 65, No. 5, 10 September 1993, pp 369-370.

The news items written by Dr R. Ramachandran and me on the impact of the cryogenic contract being renegotiated in dollar terms appeared in the *Economic Times*, Delhi edition, 12 December 1995 and *The Hindu*, 13 December 1995.

The information on the Russian cryognic engine and stage being supplied to India was got by using the facilities of the Internet:

- The information on Glavkosmos came from the web site of the Federation of American Scientists, http://www.fas.org/.

The URL of the document, titled 'Glavkosmos', which I downloaded was http://www.fas.org/spp/civil/russia/glavkosm.htm. There are other FAS documents online relating to the GSLV and Khrunichev.

- Mark Wade's online *Encyclopedia Astronautica*, http://www.friends-partners.org/~mwade/spaceflt.htm, has information on the GSLV, the KVD-1 and the RD-56M.
- Asif Siddiqi was good enough to e-mail me considerable information about the history of the KVD-1 and the IID 56 engines, much of which I have used. He also has a web site on Soviet/Russian rocket engines, http://home.earthlink.net/~cliched/engines/engine.html.
- The Proton Mission Planner's Guide can be found on the International Launch Services' web site, http://www.ilslaunch.com/. The version of the Guide which I have used is LKEB-9812-1990, Issue 1, Revision 4, 1 March 1999.

The *Performance Budget for 1995-96* is the first where the Cryogenic Upper Stage (CUS) Project is mentioned (pages 12-13). The sanctioned cost for the project is given as Rs 335.89 crore. The *Performance Budget* states:

> The objective of the Cryogenic Upper Stage (CUS) Project is to develop and qualify a restartable Cryogenic Upper Stage using about 12.5 tonnes of Liquid Oxygen (LOX) and Liquid Hydrogen (LH2) and burning for about 740 seconds with a nominal Engine Thrust of 7.5 tonnes. The Project scope also provides for delivery of two flightworthy stages for GSLV.

Cryogenic Engine for GSLV, *Propulsion Today*, Vol. VIII, No. 2, May–August 1995 gives some sketchy details about the CUS Project.

Subscale cryogenic engine:

- 'Subscale Cryogenic Engine: Water Cooled Version Tested', *Propulsion Today*, 1989.
- 'Subscale Cryogenic Engine Successfully Tested', *Propulsion Today*, Vol. VIII, No. 2. May–October 1996.

- 'Development of Subscale (One-Tonne) Cryogenic Engine Completed', *Space India*, October 1996–March 1997.

Electroforming technology:

- 'Electroforming technology', S.C. Ghosh, *Propulsion Today*, Vol. V, 1993.
- 'CECRI Expertise on Nickel Electroforming of Coolant Outer Jacket for Cryogenic Rocket Engine Thrust Chambers', *CSIR News*, 30 January 1997.

Electrical Igniter:

- 'Electrical igniter for Cryogenic Engines', *Propulsion Today*, Vol. VI, No. 3, July–September, 1994.

I reported the derated testing of the indigenous C-12 engine in 'ISRO tests indigenous cryogenic engine', *The Hindu*, 27 February 1998, p 11. The first test of the CUS engine in mid-February 2000, however, failed. The ISRO headquarters press release 'ISRO Commences Cryogenic Engine Tests', press release No. PPR:D:04:2000 dated 17 February 2000 stated:

> An important milestone in the development of indigenous Cryogenic Upper Stage for India's Geo-synchronous Satellite Launch Vehicle was achieved yesterday (16 February 2000) when the first Cryogenic Engine, employing liquid hydrogen and liquid oxygen, was ignited at Liquid Propulsion Systems Centre test complex at Mahendragiri in Tamil Nadu. However, the test had to be aborted at 15 seconds instead of the planned duration of 30 seconds. The voluminous data obtained from the test by the elaborate instrumentation during the test will be analysed to pinpoint the anomaly during the test and take suitable corrective action.
>
> The test of the first Cryogenic Engine yesterday has several accomplishments: fabrication, assembly and integration of the complete cryogenic engine; validation and commissioning of the test stand; chill down trials of the engine and associated system; production of cryogenic propellants to required specifications and; validation of appropriate safety procedures besides collection of valuable data during the 15 second test.

Since the engine was moved to test stand on 23 December 1999, elaborate trials with cryogenic propellants were carried out as part of final preparation and checks prior to the conduct of test.

Yesterday's test also marks the beginning of a series of ground qualification trials of the engine that will be carried out during the next several months.

World commercial launch market:

- *World Space Industry Survey, 10 Year Outlook*, Euroconsult, 1993 edition.
- *Le bulletin: Informations Aéronautiques et Spatiales*, 25 February 1993, No. 1571, Groupement des Industries Françaises Aéronautiques et Spatiales (GIFAS).
- 'Arianespace and telecommunications', *Arianespace Newsletter*, October 1995, No. 103.
- 'Space transportation markets', *Arianespace Newsletter*, March 1996, No. 108.
- 'The commercial space transportation market', *Arianespace Newsletter*, February 1997, No. 118.
- 'The commercial space transportation market', *e.space, Arianespace Newsletter*, April 1998, No. 131.
- 'The commercial space transportation market', *e.space, Arianespace Newsletter*, March 1999, No. 141.

Other sources of information on GSLV and cryogenic technology:

'Immediate Challenges in Cryogenic Propulsion Technology for ISRO', Keynote address by A.E. Muthunayagam, and 'Technology Problems in Cryogenic Propulsion System' by Dr E.V.S. Namboodiry and Dr M.S. Bhat, *Proceedings of the Seminar on Propulsion, Astronautical Society of India*, January 1994.

- *Indian Launch Vehicle Development*, the eighth Professor Brahm Prakash Memorial Lecture delivered by Prof. U.R. Rao, The Indian Institute of Metals, Bangalore Chapter, August 1992.
- *GSLV Bulletin*, issue No. 1 onwards.

- 'GSLV Takes Shape', *SHAR News*, No 26, October–December 1990.
- 'Gas Generator for C-12 Engine Tested', *Propulsion Today*, Vol. IV, No. 2, April–June 1992.
- 'Cryo System Facilities at Mahendragiri', *Propulsion Today*, Vol. VI, No. 1, January–March 1994.
- 'ILHP Undergoes Commissioning', *Propulsion Today*, Vol. VI, No. 3, July–September 1994.
- 'First Flight Cryogenic Stage Received in India', press release No. PPR:D:192:98 dated 23 September 1998, issued by ISRO headquarters.

Chapter 10

The Past, The Present and the Future

The analysis of ISRO by the Federation of American Scientists referred to in this chapter is titled 'Indian Space Research Organization (ISRO).' It was implemented by Christina Lindborg and downloaded from http://www.fas.org/spp/guide/india/agency.

Growth in Staff & Annual Budgets–Graph III

The budget figures were compiled from the *Performance Budgets* and the staff strength from the *Annual Reports*. For years for which I did not have data, the information was taken from Dr A. Baskaran's doctoral thesis, *Technology Development in India's Space Programme 1965-1995: The Impact of the Missile Technology Control Regime*.

The management culture:

'Management of the Indian space programme', Y.S. Rajan, *Sadhana*, Vol. 12, part 3, March 1988, pp 289-305 and reproduced in *Developments in Fluid Mechanics and Space Technology*, edited by R. Narasimha and A.P.J. Abdul Kalam, published by the Indian Academy of Sciences, 1988 is useful for understanding ISRO's management systems.

Industry as a partner:

Relevant publications have been given in the bibliography for Chapter 8, 'PSLV: Achieving Operational Launch Capability'. I have also made use of 'Space and Industry', the Shri Ram Memorial Lecture delivered by Satish Dhawan in February 1983 and reproduced in *Prof. S. Dhawan's Articles, Papers and Lectures (November 1966 to December 1994)*, published by the Indian Space Research Organization, July 1997.

The building block technique:

- 'Building Block Technique', *Countdown*, Nos. 141-144, January–April 1992.
- 'Building Block Technique for Propellant Grain', *SHAR News*, No. 40, April–September 1994.

Expendable launchers costing 10,000 times more per flight than airliners is taken from 'The prospects for European aerospace transporters—Part I: The derivation of a first order parametric method for estimating the development cost of aerospace transporters', D.M. Ashford and P.Q. Collins, *Aeronautical Journal*, January 1989. Parts II and III of their article appeared in the February and March 1989 issues respectively of the *Aeronautical Journal*.

There is a considerable amount of material available on various web sites about the US and European reusable launch vehicle programmes. Some of these web sites are:

- X-33, X-34, X-37 – http://stp.msfc.nasa.gov/
- VentureStar – http://www.venturestar.com/
- Roton – http://www.rotaryrocket.com/
- Kistler Aerospace Vehicles – http://www.kistleraerospace.com/
- The European Space Agency's Future European Space Transportation Investigations Programme (FESTIP) and Future Launchers Technologies Programme (FLTP) – http://www.esa.int/

Index